SURVIVING

SURVIVING

HOW ANIMALS ADAPT TO THEIR ENVIRONMENTS

Alessandro Minelli and Maria Pia Mannucci

FIREFLY BOOKS

CONTENTS

FOREWORD

After Alice steps through the looking glass, she enters a world in which everything is in constant motion. "In this land," the Red Queen tells her, "you must run as fast as you can just to stay in one place." The world we live in today is much like the Red Queen's kingdom, a world in which the only alternative to evolution and adaptation is extinction. Inevitable change and adaptation is obligatory for every animal and plant species, including human beings.

The underlying principles of evolutionary theory are that all living creatures are linked through common ancestry. Nonetheless, it is pretty clear that the continued existence of individual species is tied to their ability to solve problems or adopt strategies that insure survival on a daily basis. Survival strategies differ significantly even among very closely related species. As a result, the tree of life has countless branches reflective of constant change and diversification, including the branch that led to the evolution of humans. Humans, animals and plants are not vastly different, but are all part of the evolutionary process and the tree of life. The life history of every species is exceedingly complex and not easily described in simple terms. Major branches of the tree of life have divided in directions that have led to chimeras. Goethe once wrote "No living thing exists in isolation but is part of a multitude. Even we, as individuals, are part of a larger collective of living creatures." Perhaps these were just a poet's musings, but they foreshadowed what the theory of evolution was to demonstrate later on. We are indeed all chimeras. In fact each of the countless cells in our bodies is a chimera. Each is a descendant of a primeval cell, itself the result of a symbiosis of individual microbial organisms, which eventually became the internal organelles of complex cells.

Despite much effort, chimeras defy ready classification. In fact they challenge accepted rules of morphogenesis, the manner in which life forms have developed. They also raise questions as to how we categorize the natural world. The evolutionary process has given rise not only to new species like humans and *Tyrannosaurus rex*, or new classes of insects, fish or birds, but also to novel ways in which new species and classes develop. In other words, not only are the products evolving further, but the laws and rules of the evolutionary process itself are changing.

The principal focus of biological evolutions seems to hinge on origins. However, we should not delude ourselves that we have the answer just because we can demonstrate that the last common ancestor between ourselves and chimpanzees lived about five million years ago. There is an uninterrupted chain of generations between that ancestor and modern humans, each virtually indistinguishable from its predecessor and its follower. There is no sudden transition from prehistoric man to modern humans. Similarly, there is no instant in time when a common ancestor transitioned to what is today a chimpanzee.

The tree of life has many branches, each incrementally different from the others. The evolution of every species is literally based on the genealogical continuity among all living things. We cannot retrace the course of evolution nor witness every stage in human development, since species only change very gradually and within their natural environment. The evolution of a species is not like the flow of a river with a clear source and endpoint. The stream may meander backwards and forwards across the land but ultimately flows toward a clear end, the sea. The evolution of species is less like a river and more like water itself; its limits are undefined and probably unimportant. Like water, in constant motion and without confinement, every species and every individual is always changing.

And so we see that the evolutionary course of any species is dictated by the search for food, the avoidance of enemies and the adaptation to environmental changes, in a struggle both for survival and for reproduction, to insure its continuity. As a result, every life-supporting niche on Earth has been occupied, and this book is devoted to that particular aspect of survival.

HUNGER

PINPOINT

Whether active or passive, hunting delineates the border between two domains; predator and prey. In the constant struggle for survival, the hunter must rely on its skills to lure prey into its territory and then overwhelm it with a surprise attack. A small fish native to the mangroves of Southeast Asia is master at this. With a perfectly aimed stream of water, this fish easily captures flies and other small insects, which are probably more concerned about attacks from above than below.

A BAMBOO SHOOT CAN GROW MORE THAN 3 FEET OVER THE COURSE OF 3 DAYS. A FLY CAN TRANSITION FROM LARVA TO EARLY ADULTHOOD OVER A SIMILAR PERIOD OF TIME. IN THE COURSE OF SUCH EXTRAORDINARY GROWTH, BOTH THE BAMBOO SHOOT AND THE FLY LARVA GENERATE MANY NEW CELLS AND OTHER BUILDING BLOCKS. IN ADDITION TO WATER AND SMALL MOLECULES, THIS BIOLOGICAL MATERIAL CONSISTS OF VERY LARGE MOLECULES THAT ONLY LIVING ORGANISMS (AND MODERN TECHNOLOGY) CAN SYNTHESIZE. THE BUILDING BLOCKS FOR GROWTH OF EVERY ORGANISM CONSISTS OF CARBOHYDRATES, PROTEINS AND LIPIDS, AS WELL AS NUCLEIC ACIDS FOR THE SYNTHESIS OF DNA AND OTHER COMPLEX MOLECULES.

In this regard, plants enjoy a distinct autonomy that animals lack. Thanks to the process of photosynthesis, all they need is sunlight as a primary energy source. This allows them to combine inorganic molecules into the basic carbohydrates for metabolic purposes and to synthesize other organic molecules. Fungi and animals do not enjoy such autonomy; they depend totally on the energy contained in the bonds of atoms within organic molecules.

Consequently animals depend entirely on plants and other animals for their primary nutritional requirements. This naturally hinges on whether or not they are successful in their search for food. It becomes a life-long pursuit for every animal in order to survive long enough to reproduce. The need for food is particularly acute when animals are young or in their larval stage, which are times of rapid growth. Females also undergo periods of enhanced nutritional need, when they are brooding, pregnant or suckling. Nutrition is not only needed during these critical stages, because all organisms consume energy even when resting. Breathing, circulation and brain activity all depend on the production of energy, and not just minimal levels.

It is precisely because organisms depend so much on energy input, that many species have adapted to withstand extraordinarily long periods of fasting. The common field mouse, for example, survives many months of winter hibernation without a single morsel of food. That is because it feasted extensively during the preceding months to build up a reserve of fat. Even more remarkable is the capacity of migratory birds to survive extended periods of time without nutrition. They, too, rely on fat reserves, with the caveat that they must also migrate thousands of miles in the process. Although they land many times during migration, they can never count on finding food at their landing places.

As if that were not enough, small birds are among those animals that require the most nutrition. That's because of their high body temperature (a couple of degrees higher than humans), and small body mass. Even though they are covered with feathers, they lose heat. This is why they feed principally on calorie-rich food like grains, seeds or insects.

In order to meet their nutritional needs, animals must find or have access to potential food sources. This is particularly true for the active, very mobile animals like birds of prey, who spot their quarry from the air, or honeybees, who search flower after flower for pollen and nectar. Many animals do not move about much and so cannot search out food sources. In this way they behave more like plants, whose fate is almost entirely dependent on their location and environment:

water from the soil, ambient sunlight during the day and throughout the year, as well as air temperature and competition with surrounding vegetation.

Like plants, many marine animals are also sessile, including sponges and oysters. They obtain their nutrients "free" through water currents, which carry countless numbers of tiny organisms, including bacteria, single-celled algae and eggs and larvae originating from many animals. In addition, various particulates of organic matter are available to filter feeders. In this way oysters, with their specialized mouths and digestive organs, ingest various nutrients and particulates and excrete wastes. The most primitive animal forms, such as sponges, lack both mouths and excretory organs. In those cases, individual cells take up nutrients for digestion, in much the same way as ameba and other protozoas. Oral cavities and digestive organs enable small animals to feed on larger prey.

While filter feeders can survive in aquatic environments, a sessile animal on land filtering air would obviously starve to death. There would simply not be enough pollen and spores in the atmosphere to sustain them. Spiders are the only animals whose hunting strategies bear some similarity to sessile aquatic animals. They spend days without moving in the hope of snaring prey in the sticky threads of their webs. Unlike filter feeders, spiders do not consume particulate matter, but like all predators they seek large prey. However, not all spiders spin webs or lay in wait; many move about hunting for prey much as cats and birds of prey do.

Among lurking hunters the ant lion is perhaps the most unusual. It burrows under the sand with only eyes and curved mouth parts exposed to the surface. When an ant or other small insect enters the small depression in the sand, the ant lion covers it with sand and prevents its escape. The prey is then quickly grasped in the ant lion's mandibles and injected with digestive juices. The prey's internal contents are rapidly liquefied and then sucked up by the predator. In this way, the ant lion obtains a concentrated meal rich in organics and with virtually no wastes. This is important because the ant lion's gut cavity is closed and cannot be emptied during its entire larval stage. It only opens during adulthood, sparing the larva this unpleasant task.

Many lurking hunters attract their prey through deceptive maneuvers. For example, the anglerfish has a gigantic mouth covered by a fleshy growth that acts like a large lure. Some female glowworms resort to trickery: their light attracts not only potential mates but also males of related species, who run the risk of encountering a hungry opponent instead of a mate.

In situations where hunter and prey are related, they usually employ similar means of attack and defense. In general, however, attackers have a distinct advantage over their prey, be it strength, claws, teeth or other weapons. But this is not always the case. The delicate jellyfish, which preys on small fish, uses a technique that very quickly subdues its prey. Jellyfish tentacles contain numerous cells called nematocysts. Upon the slightest touch, these extend into the prey and emit potent, paralyzing venom. Only then does the jellyfish begin feeding.

Whether lured by light or chased down, the prey's fate is always the same; it is killed and consumed, although not totally. There is usually something left for other animals to scavenge.

However, it is not in everyone's best interest to destroy the creatures they feed on. What good is it for leeches or mosquitoes to have their hosts die? What is to be gained by the tapeworm or malaria carrier to endanger the lives of their hosts?

Parasites use strategies diametrically opposite to those of predators, which kill and devour. After one catch predators must pursue and catch fresh prey to survive. For parasites, the death of its host means it must seek another. That can be very difficult, particularly for parasites like tapeworms or malaria-causing protozoan, which reside inside their hosts and can neither fly like mosquitoes nor jump like fleas. For parasites it is best to live off their hosts without killing them, while severely restricting their existence. In order to propagate, parasites usually depend on some intermediary as a carrier, like the anopheles mosquito, which transmits the parasite from one person to another.

That is probably why most parasites render their host only partially incapacitated, unlike carnivores who often leave but a little meat on the bones of their prey. However, some parasites are like ticking time bombs that progressively infiltrate their host and only kill it when it is of no use anymore. That is the strategy used by many small wasp species whose larvae develop within those of other insects. The female wasp lays one or more eggs onto the host organism. The emerging larvae initially feed on the less vital parts of the host so that it can continue to feed and grow for the benefit of the wasp. Host and parasite part company the moment the wasp pupates and metamorphoses, leaving the host nothing but an empty shell.

Things are a little less spectacular with animals that feed on plants instead of other animals. To be sure we are not talking about insects that can devastate crops in the field or in grain silos. Herbivores often have no less difficulty obtaining nourishment than predators or parasites. This is mainly due to the relatively meager nutritional value provided by plants leaves, let alone stems and roots. Consequently herbivores must consume huge quantities of vegetation, even when there is no shortage of nutrients available, forcing them to spend much time in open pastures and risking exposure to predators. This is not a risk faced by insect larvae that burrow into wooden stems. On the downside, such larvae obtain only a very low energy meal that is also difficult to chew. Moreover, once embedded in a stem, the larva cannot move from its location prior to metamorphosis and is literally a prisoner in the plant in which it was hatched.

In contrast, the flesh of fruit has far more nutritional value. Only pollen and seeds are richer in nutrients, but the supply is clearly more restricted. Animals that specialize in feeding on pollen or seeds, like bees and butterflies, have developed modified mouth parts. Proper timing is important for insects that deposit their eggs into the ovaries of flowers. The developing larvae in these cases have several seeds at their disposal, rich in essential proteins, carbohydrates and fats. Unfortunate indeed is the insect that picks the wrong flower or whose timing is off when depositing its eggs.

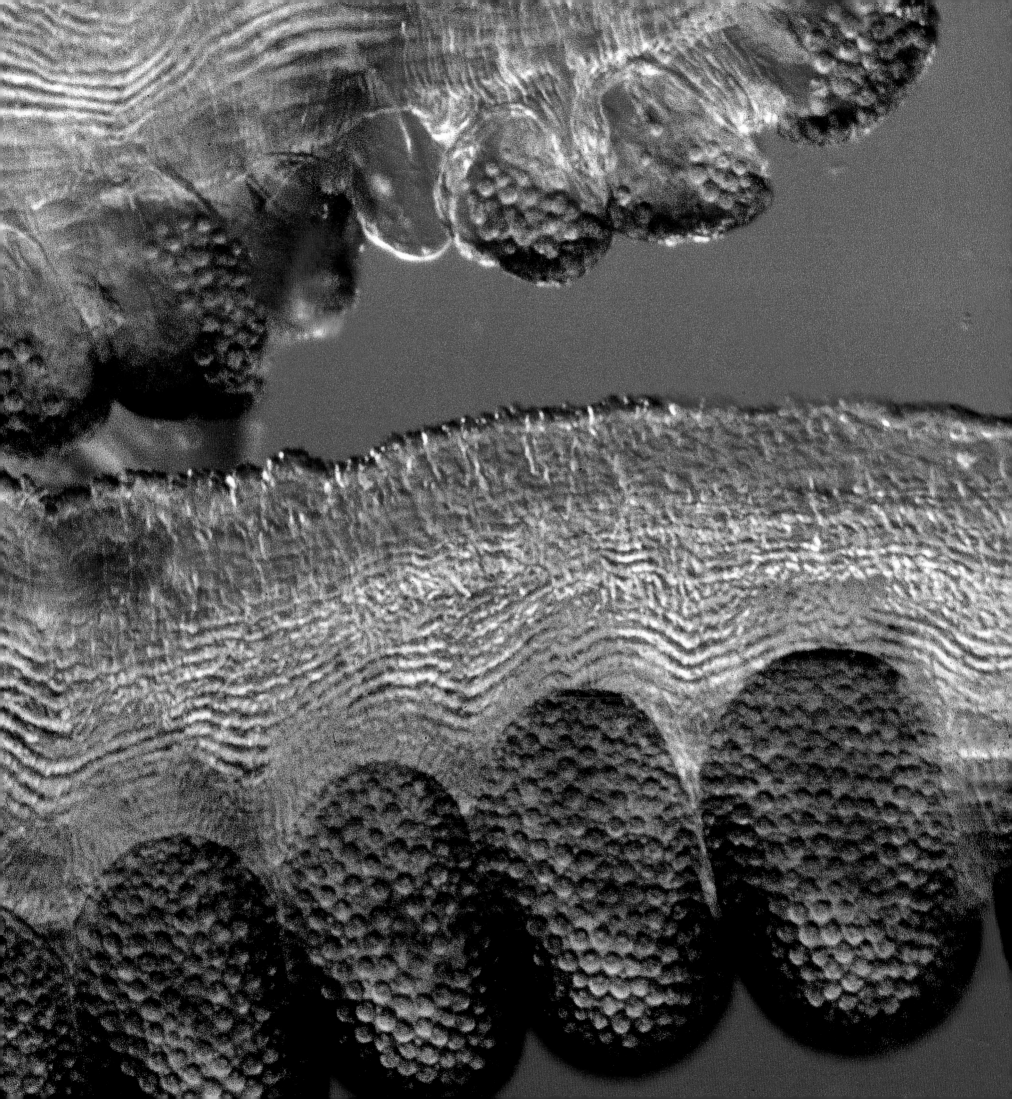

DEADLY JELLY

Thanks to its huge gas bladder, the Portuguese man of war floats on the water's surface, carried by waves and currents. Harmless as this jellyfish might appear, it is actually a dangerous creature that depends entirely on chance to capture prey. It bides its time until a careless fish comes by and gets ensnarled in the jellyfish's tentacles, and then nematocysts inject paralyzing venom into the prey.

READY TO HUNT

Predators employ two basic techniques: cunning or pursuit. The dragonfly uses both strategies in the course of its life. As a juvenile in streams and ponds, armed with formidable mandibles, it lays in wait to pounce with lightning speed on to other insect larvae, small crustaceans and tadpoles. As an adult, thanks to it powerful wings, the dragonfly hunts other insects with great speed and success. Aided by large eyes, it spots its prey from afar.

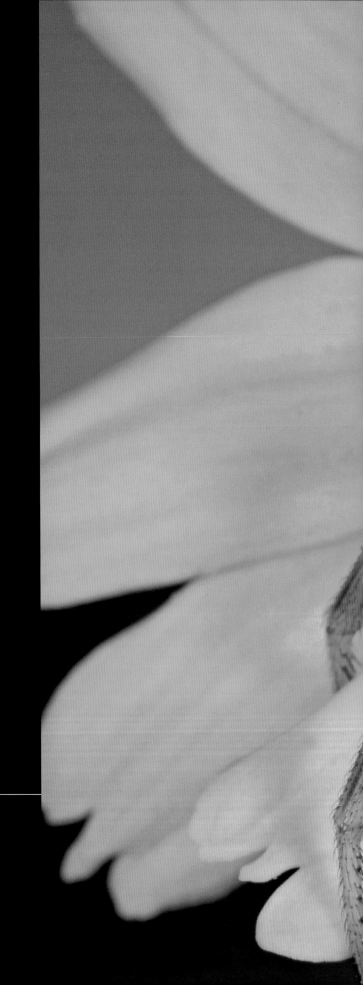

INCOGNITO

Survival-enhancing camouflage is not just found among animals trying to avoid predation. Hunters can also benefit by not being recognized immediately, even those that need not fear attacks from enemies. Hunger is always the enemy, of course. The hunter must capture prey before the end of day. Its chances of success are greatly enhanced when it lies in wait unnoticed. That is the strategy used by various crab spiders, which sit motionless within flowers and completely blend into them: red on red, white on white and yellow on yellow flowers.

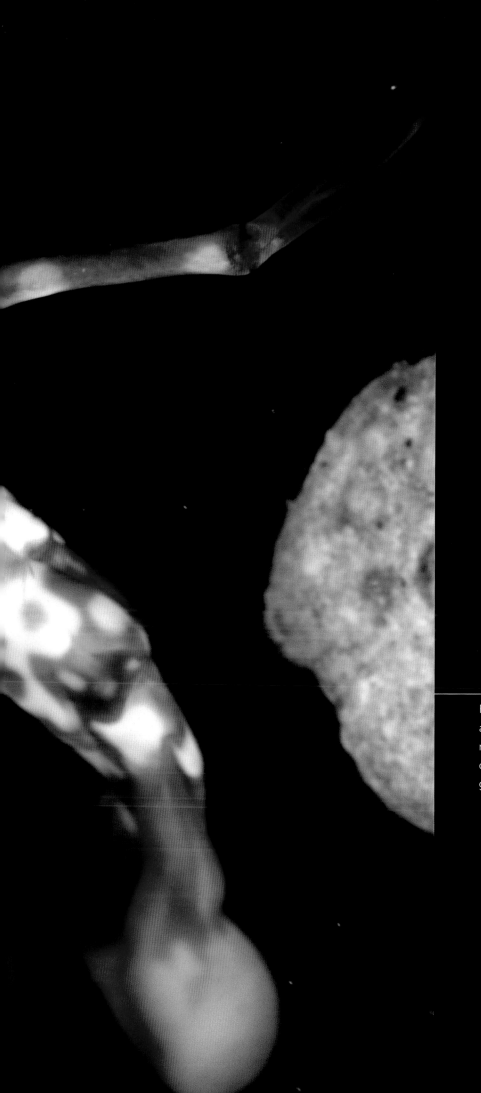

For hunters lying patiently in wait, pote[r]
again in a split second. In order to impro[v]
many predators have developed extraor[d]
couples large, protruding, multiple eyes
grasp its prey.

KILLER MOUTH

Strength and mobility are indispensable advantages for hunters that chase their prey, including orcas, whose menu includes dolphins and large fish. Such predators must also be fast enough to seize their prey in a second and thwart its escape efforts. Since their forelimbs are fins lacking digits or claws, killer whales have no option but to seize their prey and tear it apart with powerful teeth and massive jaws.

MASTER OF THE RIVER

Conventional wisdom says that big fish eat small fish. Although this is not always the case, it does happen frequently. Generally this occurs because the aggressor has strength or some other advantage over its victim. Chasing too small a prey, however, can be a waste of both time and energy. Consequently a predator may gradually change tactics over its lifespan and select prey size commensurate with its own body size.

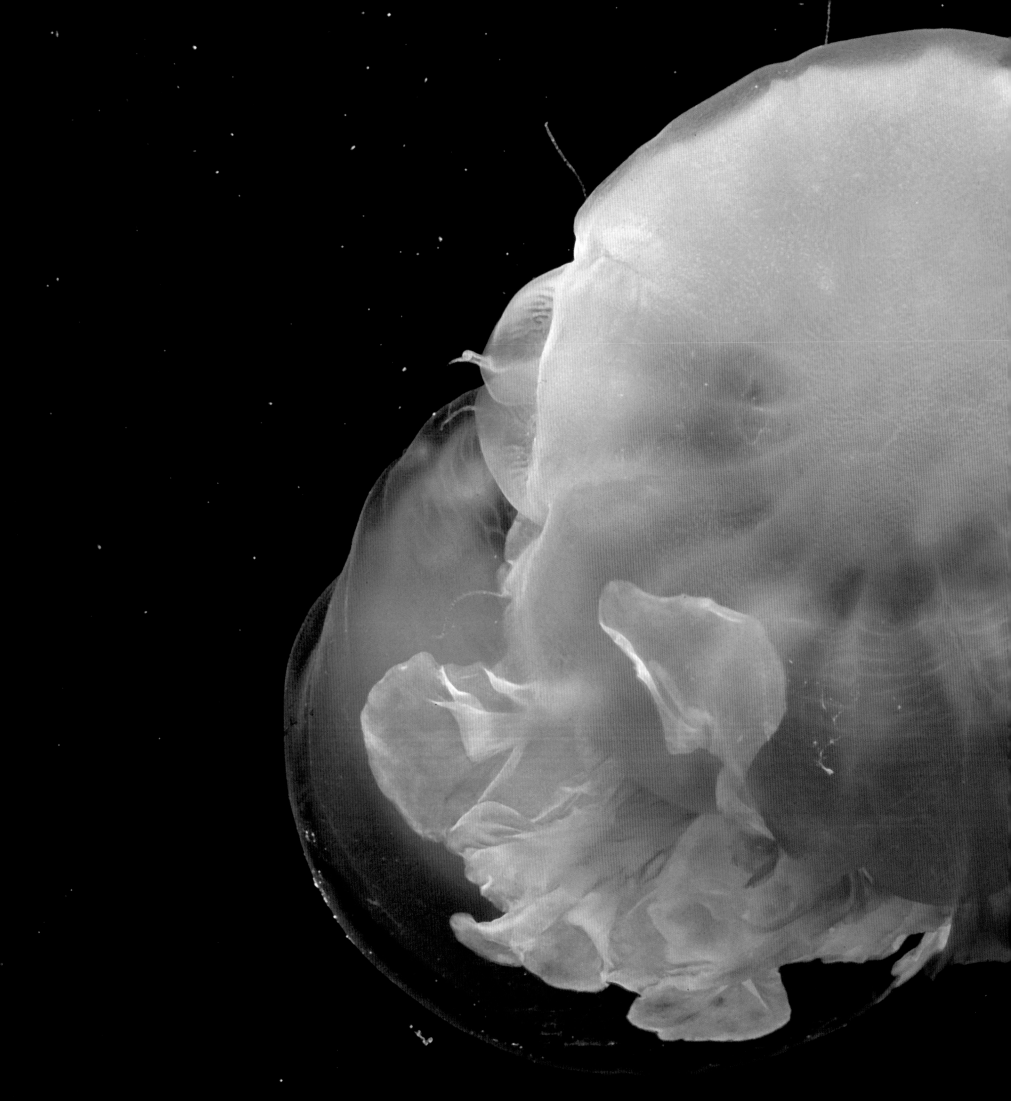

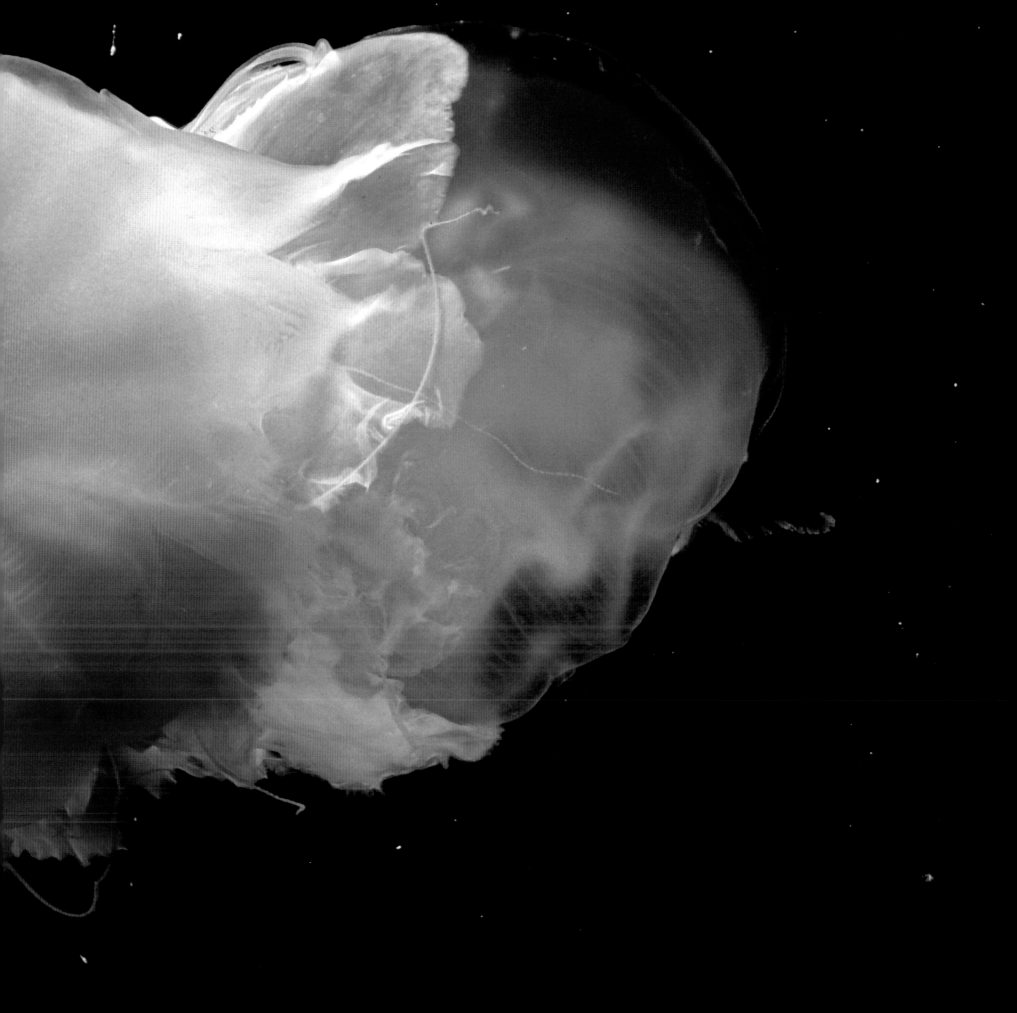

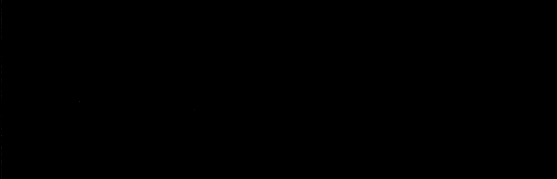

ADJUSTABLE MOUTH

The girth of a snake's prey is often larger than that of the snake itself. This is particularly true of very slender tree-climbing snakes. While the snake's stomach can readily handle such a large meal, the snake has to swallow it first. This would be a real challenge for mammals and other animals with fixed jaws. Snake jaws, however, are not fixed but expandable, allowing them to open their mouths much wider than usual.

UNDER COVER OF DARKNESS

Nocturnal activity provides two advantages for small animals: the air is cooler and more humid, and they are less likely to be spotted. Still, the cover of darkness does not guarantee complete protection since there are many predators that have also adapted to nocturnal conditions. Most hunting grounds are less active at night than in daytime. Many lizard species are nocturnal hunters, especially those of the Gecko family.

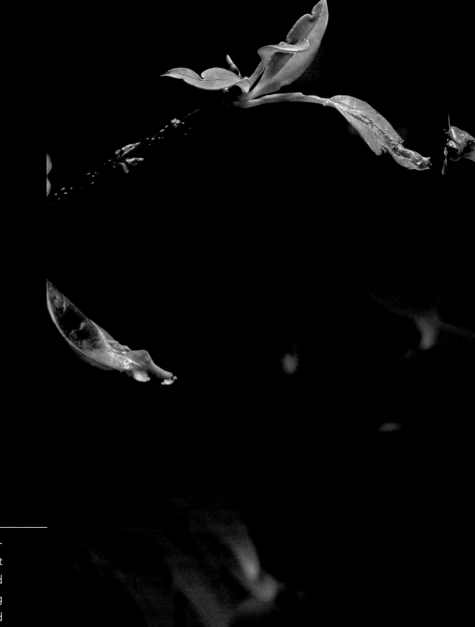

TELESCOPING TONGUE

Humans have developed many tools over the millennia, most of which serve as prosthetics: extensions of our own bodies. We are not taking about modern electronic devices, but much simpler tools like forks, scissors and baskets. For a slow and timid animal like the Chameleon, its extremely long tongue acts like a prosthetic. With its help, the animal can catch insects and immediately retract them into its mouth without having to budge.

DEADLY EMBRACE

The feeding method of the python is draining, not only because of the snake's phenomenal strength, which holds its prey in a dead-locking choke, but also because it can swallow such large prey. All of the snake's internal organs come into play, even the heart, whose muscle mass increases temporarily. Several days later, when the python can relax for a while before looking for the next meal, the heart assumes its normal dimensions once again.

AURAL EXPERTS

On their nightly food-seeking sorties, bats use an echo location system not unlike sonar. During flight bats emit a constant flow of high-frequency sound barely audible to human ears. They can detect the returning echoes from objects over a considerable distance. They can also identify flying insects this way and then capture them with open mouths while still in flight.

IRREPLACEABLE

The body temperature of birds is a couple of degrees higher that that of humans and other mammals. In order to maintain that high level of metabolism, birds require a large amount of calories on a daily basis. That is why they are in constant search of food and usually obtain their nourishment from high-energy sources like seeds or small insects. The nutritional needs of smaller birds are especially high, since they tend to lose body heat much faster.

WITH LIGHTNING SPEED

Like the reptiles that gave rise to them, ancestral birds had mouths with teeth. None of the existing avian species retain any evidence of that. Modern birds have horny beaks and use them to reduce food to small pieces. Consequently the shape of bird beaks varies considerably, adapted to their feeding habits.

FISH ANYONE?

Animals like bald eagles, sea eagles and even bears rely on surprise when they hunt for fish. The fish sense organs are of little use when they are attacked from the air. Thanks to this sharp territorial divide, birds of prey enjoy an enormously wide hunting range.

FRESH FROM THE SEA

Steep cliffs overlooking the sea offer little in the way of food, yet countless seagulls, gannets and puffins colonize them in large numbers. They feed on the abundant fish, crabs and other sea species that inhabit cold artic waters. Ironically, these bird species must spend short periods of time in the water in order to survive on land.

FORESIGHT IS BETTER THAN HINDSIGHT

In cold and temperate climates there are times of year when there is little or no food available. This poses a challenge to warm-blooded animals, which consume more energy to keep warm during the coldest times of year. Many animals migrate to warmer climates or hibernate. Planning is required to insure enough reserves are available either for hibernation or during migration. That is why many animals consume more food during warm times of year than they actually require.

ALWAYS WITHIN REACH

The biggest advantage of the giraffe's long neck is that it allows the animal to graze on the high foliage of sparsely distributed acacia trees in the African savanna. During the dry season most of the grassy vegetation in this environment is almost totally absent. For a couple of months, the green tree foliage provides welcome nourishment. That is assuming of course that the giraffes can find it.

COMMUNAL MEALS

Not many predatory animals live together in groups, but it does make hunting easier and then they can feed communally on the downed prey. Tigers and leopards, eagles and falcons, dragonflies and bats all hunt solo. Cheetahs and wolves are exceptions, as are female lions; they hunt in groups and then feed in order of dominance.

MILK

Unlike many animals that lay several thousands, sometimes even millions of eggs, mammals bring far fewer offspring into the world. Certainly pregnancy is an extremely draining period. Newborn mammals require considerable care, often even from male parents, particularly when suckling. Parental care can extend beyond weaning and quite a bit longer.

FOR SINGLE-CELLED ORGANISMS, LIKE BACTERIA OR AMEBAS, REPRODUCTION IS ASSURED THROUGH THE PRECISE MECHANISM OF CELL DIVISION. EVERY METABOLICALLY ACTIVE CELL CAN UNDERGO REPLICATION AND DIVISION, LEAVING EACH DAUGHTER CELL A PRECISE COPY OF ITS GENES AND A FULL COMPLEMENT OF MOLECULES AND ORGANELLES. THE NEW CELLS CAN THEN IMMEDIATELY RESUME NORMAL GROWTH, UNTIL THEY, TOO, ARE READY TO DIVIDE. THAT ASSUMES THEY SURVIVE THAT LONG AND ARE NOT CONSUMED OR DESICCATED BEFORE THEN. IF THIS DOESN'T HAPPEN, SINGLE CELLS ARE ESSENTIALLY IMMORTAL. WHEN AN INDIVIDUAL MOTHER CELL DIVIDES AND GIVES RISE TO TWO DAUGHTER CELLS, NONE OF ITS ORIGINAL COMPONENTS ARE LOST BUT ARE EQUALLY ALLOCATED TO ITS TWO DAUGHTER CELLS.

Cell division therefore, seems to assure the continuity of life through proliferation of new cells, replacing any that are lost from the biosphere. The reality, however, is far more complex. For many millions of years life on our planet consisted entirely of single-celled forms, but about a billion years ago the first multi-celled organisms made their appearance and gave rise to fungi, plants and animals.

Two possible mechanisms likely gave rise to multi-cellular organisms. One option is that originally independent single-celled forms began to form aggregates or associations. This did create a risk, whereby some cells in the aggregate might dominate others to further their own reproductive advantage. Consequently this type of cell aggregation is found primarily among amoeboid forms, which swarm into larger structures as a launch bed for sporulation. In all other cases, whether plant or animal, multi-cellular association formed through cell division without subsequent cell separation.

In animals, the entire process of embryonic development can be followed step by step after the first division of the fertilized egg cell. The latter is among the largest cells in animals and its cleavage proceeds from two cells, through four, eight, and so on. One advantage of this process is that it insures continuity of species, plant or animal, because they can reach sizes far larger than any single-celled organism could.

Although much larger, multi-celled organisms have only limited life spans. Only in rare cases can multi-cellular organisms survive and reproduce by simply dividing into two parts like single-celled species. Most have developed other reproductive strategies.

The reproductive process in animals almost always starts with a single cell with the potential to generate a multi-celled organism. This single egg cell, the ancestress of an eventual multi-celled organism, contains an extraordinary supply of nutrients sufficient for the step-by-step development of the embryo. There are enough nutrients to bridge the time required (which can range from a day to several months, depending on the species) to ensure the development and survival of the offspring until it can fend on its own. That is why eggs are very large cells whose bulk consists of yolk to nourish the embryo.

The egg cell with its nutrient supply is not sufficient to insure future generations. Left alone the egg will eventually deteriorate. The egg's natural partner is another cell, the sperm, which must fertilize the egg in order to generate a new individual.

The principles of sexuality come into play during reproduction. The process brings together hereditary material from two different cells, each of which has undergone chromosomal recombination to generate variations that might prove advantageous to future generations.

In an earthly environment that is constantly changing, there is no final or perfect adaptation. To survive under such circumstances, it is best not to adopt one long-perfected strategy, since entirely new requirements may be needed in the future. One mechanism to generate random changes is through combinations of genetically different germ cells in sexual reproduction.

The importance of reproduction in multi-celled organisms consequently does not rest solely on production and release of single cells with the properties of their mother cell. Such cells must also be able to unite and generate a significant number of recombinants, only some of which survive, either by chance or due to their exceptional characteristics. Survival can be a thorny issue for germ cells and the individuals producing them. Egg cells and sperm, which are usually produced by different individuals, each carry unique genetic traits that are not likely to occur again. In many species therefore, the strategy is for females to produce as many eggs as possible, since they have limited mobility or are totally sessile. Males of the same species, who usually produce and carry much smaller sperm cells, use their mobility to seek out mates and deposit their highly mobile sperm.

In addition to this division of labor between sexes with respect to egg and sperm production, specialized communication strategies are also required. Such strategies are particularly common among animals that are sessile or have only limited mobility: notably corals, oysters and jellyfish. These animals simply release their eggs and sperm into the surrounding water, where chances of successful fertilization are essentially random. Reproduction cycles are tied to lunar phase cycles to insure that maximum egg and sperm releases in a given population occur in similar time periods. Successful fertilization would still be unlikely however, were it not for the fact that the egg cells release a specific chemical (pheromone) into the water to help sperm orient in the direction of their intended goal.

External fertilization can be the only opportunity for a sedentary aquatic animal to secure progeny, but this is a very hit-and-miss situation, given the near uncountable numbers of germ cells that are produced.

For mobile animals a totally different mode of reproduction exists, namely the internal fertilization of the egg by sperm, which usually takes place within the female's body. This alternative is pretty much mandatory on land, as external fertilization is effectively out of the question. External fertilization in a river, for example, would be impossible, not to mention the problem the tiny reproductive cell would face in order to survive in a saltwater environment.

The situation is less complicated for germ cells on land than for the animal that produces them, since the onus is on the animal to find a suitable partner. It is precisely for these reasons that extraordinary methods of communication have evolved in both the plant and animal worlds. Specific coloration and scent patterns have evolved in plants to attract pollinators like insects and birds, who can transport the appropriate pollen from flower to flower.

It is not enough, however, to focus on germ cell production. Another important factor is how an animal can insure its own growth. The important considerations are where to allocate resources: whether for chasing prey, fighting rivals or evading prey. This essentially means mustering the energy and taking the risk of interacting with another member of its species by exposing obvious coloration, or emitting sounds to attract a mate, which can also bring the attention of an unwelcome rival.

Some animals only pair once during their lives. Not being prepared for that opportunity means the possibility of procreation is gone for good, and there is no chance of living on, even if only indirectly, in the next generation.

Furthermore, the job is not done just because successful fertilization has taken place. The struggle to insure a future generation has only just begun for a bird, which must still lay its eggs and feed its chicks, or a mammal, which must deliver the babies and then nurse its young.

Not all male animals are solely driven by the desire to propagate their own genes; some actually participate with their partner in parenting duties until their young become independent. Male sea horses carry the eggs in a pouch and are their sole guardians. Not only do some female animals participate in fostering the next generation by producing many eggs, but others purposefully try to keep their broods small. There are animals that only have a single mother, in the founding, genetic sense of the word, and no father, as well as some that have two parents, neither one of which is sexually defined.

Reproduction without a father is the result of parthenogenesis through unfertilized eggs, which is found among relatively few species, and only among female-only generated offspring, where no males are born for considerable periods of time. Although this phenomenon is very common among invertebrates, it also occurs among lizards and salamanders. This also happens among leaf aphids, where a parthogenetic generation alternates yearly with a sexually reproduced generation. In other cases, such as bees, fertilization does happen but with different results; fertilized eggs give rise to females while unfertilized eggs produce only males.

Finally we also find hermaphrodism among animals, where individuals are produced that have both independent male and female characteristics. For example, in tapeworms, it is possible to find a hermaphroditically produced egg that is fertilized by sperm from the same individual. Among snails and leeches, hermaphrodites are fertilized by other individuals. This is obviously necessary in cases where animals undergo sex changes partway through their lives, like the Squirefish, which after its initial maturation is a male, and then becomes female during its reproductive years.

Sexual characteristics are genetically determined, the result of specific combinations of chromosomes when the fertilized egg becomes a zygote. This is not always the case, however.

In many reptiles, the sexual differentiation of a given individual is linked to the temperature during the incubation time of the egg. In alligators, for instance, males develop below a certain temperature and females above it. It has been difficult to determine what advantages this type of sex determination might have, since it is so dependent on environmental factors.

In order to maintain a balance between sexes, the marine worm *Bonellia viridis* has evolved a far more secure method of reproduction that involves a miniscule male living as a guest inside the body of the much larger female. Before settling to the sea floor and turning into a worm, the *Bonellia* larva floats in the water for several days, during which time its sex remains undetermined. This does not happen until metamorphosis. If this takes place in the vicinity of a female, the larva absorbs a specific substance emitted by the adult worm. Under these circumstances, the larva develops into a male and nestles itself into the body of the adult female worm. If the larva does not absorb this substance from a nearby female at the proper time, it develops into a female which in turn, will trigger other larvae to become males.

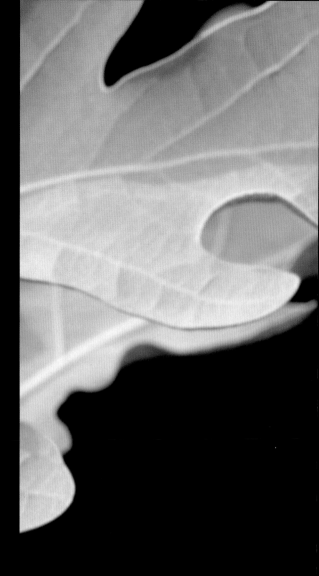

CHOOSE ME!

In a world of aggressive competitors, the male peacock must feel pretty confident to display his full plumage to everyone around. This way, he hopes to attract the attention of a female and mate with her. It's a dangerous game but one well worth the risk, since only the most attractive males procreate success-fully. In fact he succeeds in acquiring a full harem, while other males remain mate-less, at least until the next pairing ritual.

LOVE WARS

Stag beetles are among the stateliest of insects. The male's lower mandible serve a similar function to the elk's head rack, namely as weapons in the battle for females during mating season. As with stags, this beetle's namesake, fighting is largely ritualistic and generally bloodless as the vanquished simply leaves the scene.

WE BELONG TOGETHER

The eggs deposited by toads, frogs and tree frogs would not survive for long if the females left them unattended on the ground, where the gelatinous eggs would quickly desiccate. Fortunately the eggs are usually deposited in suitable locations, though not always in ponds or lakes. In tropical rain forests many frog species deposit their eggs in water bladders, commonly found on the leaves of many trees. There is even one South American species where the male foregoes feeding for several days and holds the eggs in its mouth until they hatch.

AMOROUS APES

When a male chimp first approaches a female he is anything but hopeful, but sooner or later they will pair off and leave their group for a honeymoon. This can last up to a month. Upon their return from the forest, the female is usually happily pregnant.

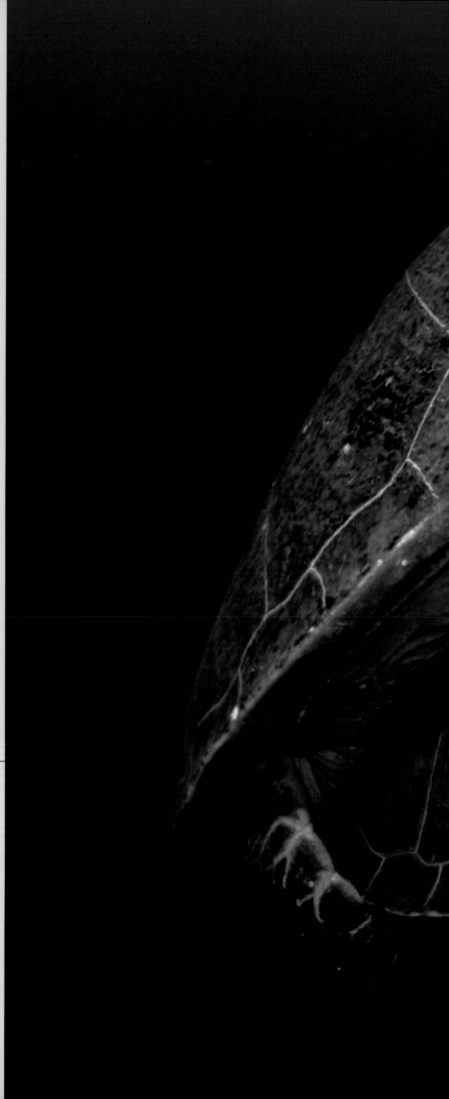

A TIME FOR TENDERNESS

The amniotic egg is one of the most remarkable achievements of vertebrate evolution. Since they are covered with a protective shell, the eggs of lizards, snakes and birds can be deposited in dry locations. Inside, the embryo has everything it needs, yolk and water. Before the egg is coated with a shell, it must be fertilized inside the female's body. This is also the case with former terrestrial vertebrates that have returned to water, such as dolphins, whales and giant sea turtles.

MOTHER KNOWS BEST

It's best to pay attention when you are young, particularly when your paren
or at least your mother is your life role model. Once you are on your own yo
must practice what you have learned.

THE WAY TO A WOMAN'S HEART
IS THROUGH HER STOMACH

Pairing up can be dangerous, especially among thieves. This is particularly true when males are smaller or weaker than their mating partners. Occasionally the female praying mantis will devour its partner while mating. As well as inseminating her with sperm, the male provides his body as a hearty meal for the female, who then produces dozens of eggs.

YOU LOOK LIKE MY TYPE

The life of most insects is regulated through chemistry. Their responsive sense organs help them find food sources as well as others of their species and mating partners. Butterflies have excellent eyes and thanks to their strikingly colorful wings, which often differ among males and females, they can attract potential partners even at considerable distances. Whether these are also willing is determined only later when the two come in contact.

STRENGTH IN NUMBERS

Mats of plankton that float on the ocean surface consist of many small algae and dozens of different crustaceans, as well as countless numbers of larvae that eventually develop into sponges, mussels, starfish and jellyfish. Chances of individual survival in this mix are slim, but their collective survival is assured. Other sea creatures, including squid and octopus, skip the larval stage completely and fend actively for themselves as soon as they have hatched from their eggs.

DADDY'S BOY

Among many animals, the care and nurture of the young are the female's responsibility. Sometimes both partners are involved, particularly when it comes to nest building, which is often the male's domain. In some species, however, the female only lays the eggs, and everything else is done by the male. This is the case with seahorses. The eggs are collected within the male's mid-pouch, and after the embryos are hatched, the father continues to care for his offspring.

ONE ENCHANTED EVENING

Many species, especially small animals, have short life spans. Because of that, they often have very brief mating periods, sometimes lasting less than a couple of hours. Some larger animals, including squid, octopi and calamari, are similarly constrained. They release their germ cells into the water in a convulsive fashion and then die a few hours later. Often many thousands of calamari assemble in a tranquil area of the sea and go through this ritual as a group.

CHILDREN OF WINTER

There is no vegetation for nest building and no shelter from the icy polar winds — only frozen desert as far as the eye can see. The only place for the Emperor penguin to brood its big egg, is in its own plumage. The male penguin carries out this difficult task. He cannot leave his position during the entire Antarctic winter. He must fast until the chick hatches during the short spring, and only then is he free to take the long-awaited dive into the sea to hunt for food.

ONE FOR MOM AND ONE FOR DAD...

Among animals that inhabit climatic zones with sharp seasonal differences, mating is timed so that offspring are born at times when food sources are at their most plentiful. That is why many songbirds seek out partners, start nest building and mate during early spring. This way eggs are hatched in late spring and the parents go in search of insects and other small animals to feed their ever-hungry young.

JUST THE RIGHT SIZE

Females are often larger than males among egg-bearing animals. The number, size and weight of eggs that females must
carry can be enormous, in contrast to the far less onerous task of
sperm production by males. Among mammals and some other
species, the relative size difference betweten males and females is reversed. Here males are usually considerably larger than
females, mainly because they are in constant competition with
other males for the right to rule a harem of the smaller and less
aggressive females.

COME PLAY WITH ME

Social animals do not just pass their chromosomes to their
offspring, their cultural traits are also handed down from one
generation to the next. Many birds learn their specific songs by
hearing their parents vocalize all day long. Predatory animals
teach their young during the prolonged periods of time they
spend together, often long after weaning. Learning to hunt
starts as playful behavior among the young and only later becomes a deadly serious undertaking.

STICKING CLOSE TO MOM

It's probably no coincidence that many large mammals only bear one off-spring per mating season. Carrying a calf, young elk or baby elephant cannot be an easy task for the expectant mother. Their young are comparatively independent only a few hours after birth; once they are on their feet, they are ready to follow their mothers and the rest of the herd from one grazing spot to the next. Apart from suckling, which is always a busy undertaking, the bonds between mother and young are not particularly strong, even shortly after birth. The situation is quite opposite among mammals like bears, whose newborn are very small and not ready to be on their own. In fact they are relatively slow in their development and totally dependent on their mother for quite some time.

IS IT A BOY OR A GIRL?

Among humans and other mammals, as well as birds, frogs and insects, the sex of the future offspring is determined the moment egg and sperm combine. This is not the case with many reptiles, however, where sex determination after fertilization remains undetermined for several days and is largely temperature dependent. In alligators, for example, females develop preferentially at relatively high temperatures and males at lower temperatures.

IT'S A GREEN WORLD

The green color of leaves, which is actually the green of chlorophyll, dominates almost every animal habitat. One tactic that will enhance a creature's chances of survival is to become nearly invisible by blending in with this green background, especially if they also have the ability to remain motionless. Geckos are masters at this. They also have another unusual capability thanks to millions of tiny adhesive structures (known as spatulas) on the base of their feet, geckos can literally walk upside down on glass surfaces.

THERE IS PROBABLY NO SPECIES OF ANIMAL THAT IS NOT AT RISK FROM SOME KIND OF ATTACK, NOT EVEN THE LARGEST OR WILDEST ANIMALS, EVEN IF SUCH ATTACKS ARE FROM AGGRESSIVE PARASITES. EVEN PLANTS, ROOTED IN PLACE WITH NO MEANS OF ESCAPE, ARE SUBJECT TO ATTACKS BY INSECTS. WHILE MANY INSECTS MAY STEAL NECTAR AND POLLEN FROM PLANTS, THEY RECIPROCATE BY CARRYING POLLEN FROM ONE FLOWER TO THE NEXT, THEREBY PROMOTING POLLINATION AND FRUITING. ON THE OTHER HAND, ATTACKS BY CATERPILLARS AND WOOD-FEEDING INSECTS POSE RISKS TO PLANTS WITH FAR MORE DIRE CONSEQUENCES.

Often offense is the best defense. Large thorns are effective protection against plant-eating animals, who risk getting their mouths cut or noses bloodied. Pungent odors help protect plants against animals interested in dining on fresh fruit or tender green leaves. Minute surface hairs can discourage plant-eating insects from lighting on leaves.

Probably the most effective defense that some plants have at their disposal is poison. This is no guarantee of success, however. Plants like the deadly nightshade, which is in the same family as potatoes, contain high levels of alkaloids. Likewise, the green parts of many legumes are rich in organic compounds that can release highly toxic cyanide ions. Such fiber-derived toxins keep a significant portion of large and small herbivores at bay, but there are several insect species that are immune to them. Unlike the majority of plant eaters, they have the molecular defenses needed to neutralize such toxins or to act as antidotes. These toxin-resistant insects have the additional advantage of not having to compete with other herbivores.

Gaps in the defense systems of these plants show clearly that no natural defense exists that can't be breached. In fact the opposite is usually true. In the constant struggle for survival, some insects have overcome the plant's toxic defenses and are able to feed safely on them. This provides these insects a tremendous advantage over competitors, as well as an additional benefit: by ingesting and incorporating the toxins into their own bodies, they are now protected against predation by other species.

When a predator dies because it has fed on poisoned prey, scavengers feeding on it may encounter a similar fate. Neither predator nor prey has gained anything. The situation is quite different if a predator can differentiate between poisoned and non-poisoned prey. Moreover, if a predator experiences only limited ill effects when eating poisoned prey and survives, it is likely to remember that and avoid feeding on poisoned prey again.

Some animals whose meat is distasteful have evolved stark body coloration, such as red and black. Animals with threatening spikes or prominent poison teeth are also readily apparent to potential predators. Wasps, with their prominent yellow-black stripes, are good examples of this, as are coral snakes, whose long bodies are covered with black and white stripes against a stark red background. Anyone who has ever been stung by a wasp will be sure to stay clear of its poison-tipped stinger in future. The wasp's prominent body coloration is beneficial to both parties. It serves as a warning to any potential attacker to stay clear. The wasp in turn does not have to hide from an aggressor and has its stinger for defense.

Poison butterflies also count on being recognized by potential predators. Although they carry toxins in their bodies in the form of larvae, without the telltale warning colors, the poison would not be evident until after they had actually been eaten by an aggressor. Animals like these clearly benefit from distinct external signals that warn their predators that they are poisonous.

In the desert, where the extreme heat forces predator and prey alike to restrict their activity to night time, striking external coloration would not necessarily help as a recognition signal. Consequently, other means of communication have evolved, such as the unmistakable noise of the rattlesnake.

Unlike wasps and coral snakes, whose stripes have been widely imitated by other species, rattlesnake tails have not been copied. Whether insect or snake, not all mimicking species have poisonous stingers or fangs: many non-poisonous species have unashamedly adopted the distinctive outer markings of their poisonous counterparts. Any creature familiar with coral snakes is likely to avoid any snake that has similar body coloration and stripes, since to do otherwise is simply too risky. The same is true for the colorful root maggot fly, which resembles a poisonous insect but is actually harmless.

The "deceit" of mimicry has its limits. When the number of wasp-imitating flies is low, chances that a potential predator will get stung are comparatively high. On the other hand, if the yellow-black striped flies are numerous, predators are less likely to associate their coloration with danger and proceed to hunt them, even though they might experience an occasional encounter with a stinging wasp.

Examples like these illustrate clearly the importance of specific signals that an animal might send to fend off predation, including body coloration, movement and sound. This will obviously only work in species that have such defenses at their disposal. Those that do not must resort to other strategies.

One such strategy, albeit not an easy one, is to avoid detection altogether. Confined as it is to living underground, the mole would seem like a perfect example. It's not absolutely necessary to dig a hole, however, to avoid being seen. Many animals achieve this end and escape predation by blending into their surroundings through coloration; green for animals living among leaves; ocher for animals roaming through dry, grassy terrain; grey or mottled coloration for those living close to the ground.

Some animals avoid detection by remaining motionless for hours on end. Even if that is its only defense, such an animal can survive by blending totally with its surroundings. Some insects look like sticks, leaves or thorns, while certain caterpillars and butterflies look like bird droppings.

Mimicry and deception maneuvers like these depend on what signals are communicated between potential prey and predator. The aim of the prey is to deceive or distract the predator long enough to escape from it. For example, several butterfly species have hind wings that look like small tails marked with prominent colored dots. When the butterfly lands on a flower and closes its wings, they resemble antennae and the round dots look like eyes, while the true head and real antennae are actually at the other end. As a result, a would-be predator is tricked into thinking the butterfly will fly off in the wrong direction and directly into its mouth. Wrong assumption!

The key aspects of any communications system, whether honest or deceptive, only work when the receiver is equipped to capture and interpret the signal. In this regard, most successful hunters have the necessary equipment: witness the keen vision of an eagle, which can spot a mouse on the ground from great heights, or a cat, whose exceptional sense of smell can detect its prey from afar.

Quite often, predators give their presence away quite unintentionally. It pays, therefore, for any potential prey to have keen vision and heightened sensitivity to enhance its chances of escape.

In this regard, the eyes of the Eurasian woodcock are extraordinary well adapted. Located laterally on the bird's small, narrow head, they allow the animal to see in all directions at once without having to move. Its field of vision is so wide that it can actually see backwards over its own shoulder, precisely the direction from which an attack is most likely to occur.

No less effective, although quite different, are the eyes of scallops. Equipped with a series of small, dark eyes, the scallop has a wide field of view when its shell is partly open. This gives it plenty of time to react to a slow-moving predator like a starfish, its most common enemy.

Of all defense mechanisms that rely on communication between predator and prey, few are more unusual than those deployed by moths to evade their primary foe, the bat. While in flight, bats emit a constant, high-frequency sound pattern that is inaudible to the human ear. It rebounds off all objects in its path, and the returning echoes are detected by the bat. Like sonar, this echolocation system allows the bat to soar in total darkness and detect flying insects, that it can capture on the fly.

To combat this type of attack, moths have developed an effective countermeasure. Since the moth can also detect the bat's ultrasound signals, it will immediately close its wings and fall toward the ground. The simple act effectively evades the bat's sonar, and thereby deters it from further pursuit.

As humans we tend to be more impressed by these high-tech evasive measures than by the armadillo's protective armor, the turtle's carapace or the mussel's shell. Some animals look for movable shelter or protection not through parts or extensions of their own bodies but in something totally external. One example of this is the caddis fly, a small, butterfly-like insect, with hairy wings and a pale coloration. Its larvae reside in water in self-built baskets made of leaves, twigs, pebbles and even small mussel shells. These materials are carefully selected, assembled and knitted together with thin threads of silk. The larvae leave their cylindrical hideouts only when they are hungry and can move about. At that point they are awkwardly exposed and subject to attacks by a predator.

Lastly, let's look at another example of a complex self-defense mechanism, that of the bombardier beetle. This small ground beetle, which lives amid the soil in humid environments, has special glands in its abdomen that contain hydrogen-peroxide and hydroquinone. These two chemicals form an explosive mixture when combined, releasing a hot, foul-smelling gas-liquid mix. When the beetle is attacked, it expels this through an outlet valve with a loud popping sound. This defense mechanism effectively deters an aggressor by exposing it to a very high dose of benzoquinone.

HIDING IN PLAIN SIGHT

Being herbivorous has its advantages. Unlike carnivores, which have to first catch their meal, grazers find plants all around them, firmly rooted in the soil. The same holds true for sea spiders, which feed on tiny polyps that grow on rock surfaces. But this is a two- edged sword: whether nibbling on plants or squid, the respective grazers expose themselves to other hungry preda- tors. Although this is a constant danger in every environment, it is less risky if the grazer's coloration blends in with that of its surroundings.

IN DISGUISE

It's not only because of coloration that many animals blend into their surroundings, to the point of being nearly invisible. This type of disguise is particularly well developed among many insects, whose body surfaces may not only resemble leaves, twigs and thorns, but also rough surfaces and folds. In this regard, the insects belonging to the genus *Extatosoma*, also known as Phamisches, have an especially wide repertoire. Those species with particularly long and narrow bodies are called walking sticks, and those with flattened bodies are known as walking leaves.

SEEK AND YOU SHALL FIND

Only a seasoned reef hunter will recognize a longnose hawkfish as it moves among the corals. As long as this bright, red-colored fish stays in its niche, it will remain invisible to anyone else. Sooner or later it will have to leave this safe refuge and lose the advantage of camouflage. The same is true for most species, as there is no such thing as perfect adaptation. Life is always fraught with danger.

LOOK INTO MY EYES

At first glance, this small South American frog looks large, black and sparkling. It's easy to confuse the large dark spots on its rear end with a pair of eyes. In the dim light of the underbrush, these large unmoving "eyes" might easily be mistaken for those of a snake. To add to its advantage, this frog can leap with lightning speed in the opposite direction from what would be expected by a predator fooled by the snake "eyes."

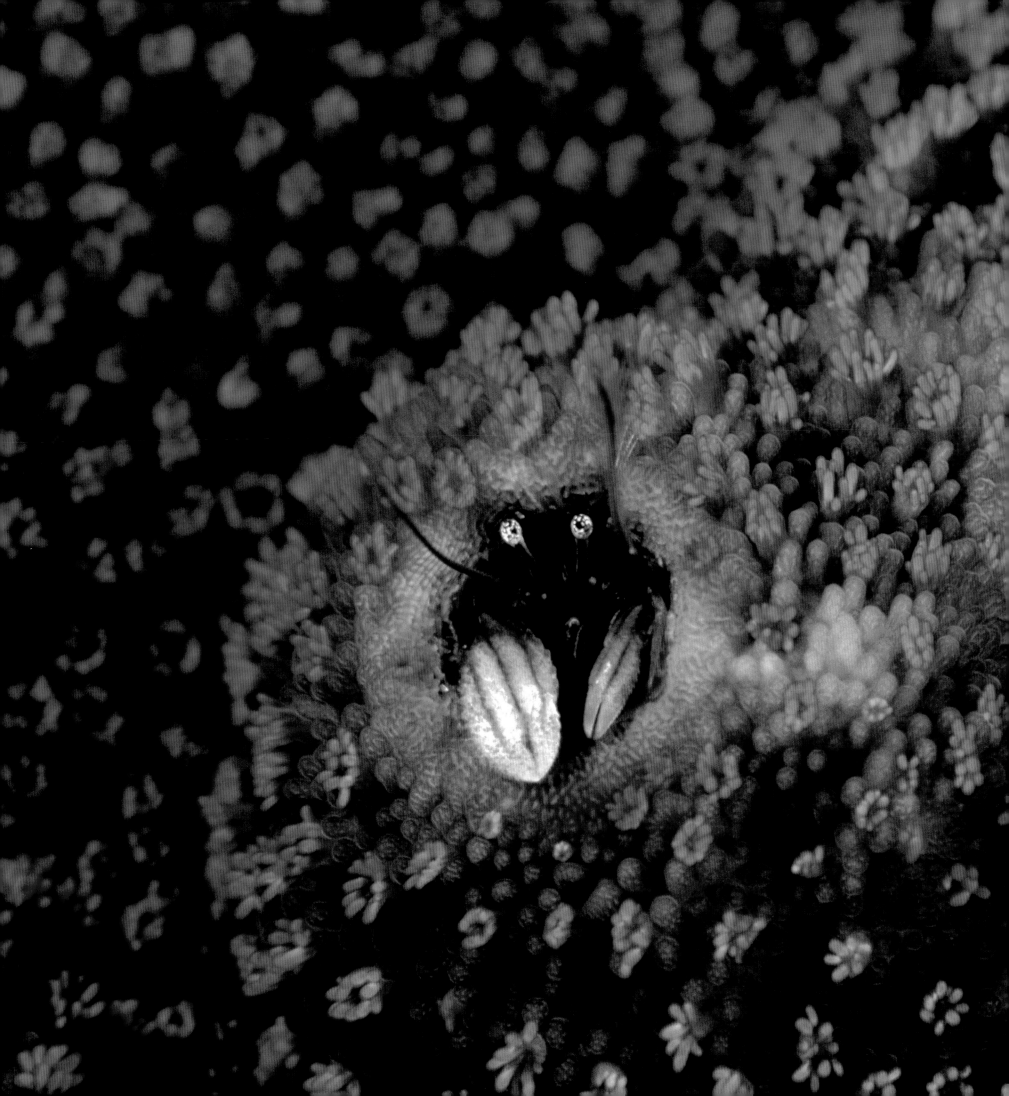

COZY LITTLE HOME

Snails and hermit crabs have one or two things in common. Both have asymmetrically shaped bodies, which in the case of snails are embedded in a spiral-shaped shell for protection. Hermit crabs (which are distantly related to true crabs) seek refuge for their distorted bodies in abandoned snail shells they find along the sea shore. Some species of hermit crabs, like the one pictured here (*Paguritta vittate*), have symmetric bodies that could not be accommodated by a spiral snail shell. Instead, they inhabit the cylindrical tubes left behind by worms, which are often overgrown with colorful corrals.

CLOWNING AROUND

Small fish that inhabit coral reefs are not just threatened by larger fish, but are also at risk of being paralyzed by large poisonous sea anemones. It's best to move with extra caution among the anemone's tentacles. Thanks to a gelatinous outer coating, clownfish are protected against being stung and can move safely about in this environment, where they are protected from predators who need to stay clear of the deadly plant-like anemones.

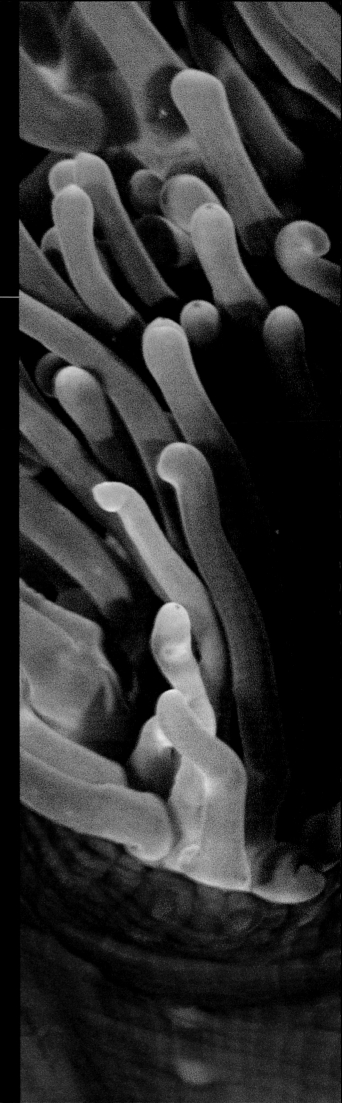

IN SEARCH OF SHELTER

Many animals are protected by natural armor, including crabs, snails, turtles and armadillos. Others, like caddis-fly larvae, have to build their own. These small, aquatic insects, which live on land as adults and resemble small butterflies with hairy wings, search rivers and ponds for building materials. Every caddis-fly aspecies uses different methods and materials to accomplish this, but they all use their own silk to bundle things together, the same silk they used to build cocoons in preparation for metamorphosis.

THE BEGINNING

The very beginning of life is often the most dangerous time for many animals. The egg or larva stage, when the animal has no natural defenses, provides easy pickings for many predators. Things are easier for species with large, yolk-rich eggs, which are deposited in calm and secluded locations. When such eggs hatch, the hatchlings are usually not larvae, but closer to full grown and more prepared to fend for themselves.

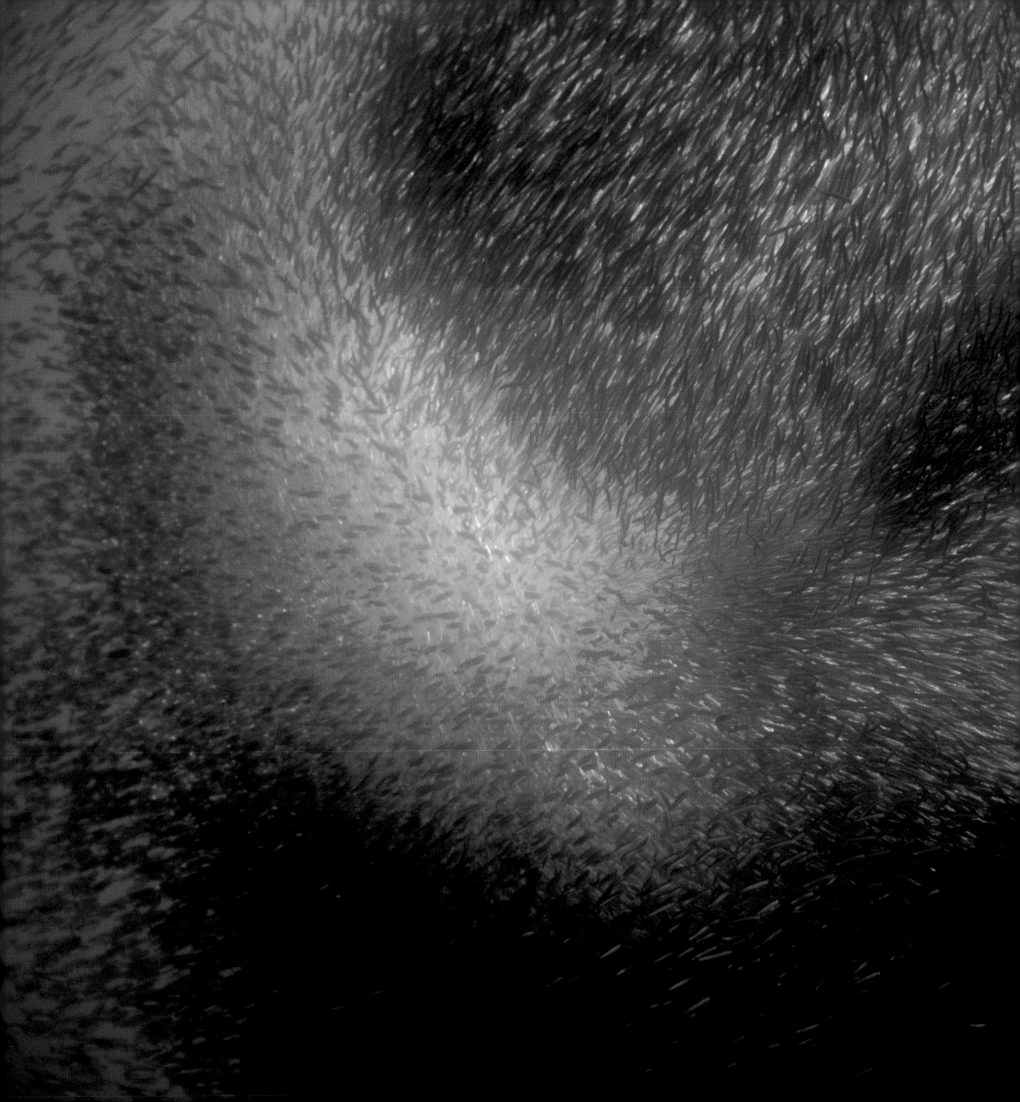

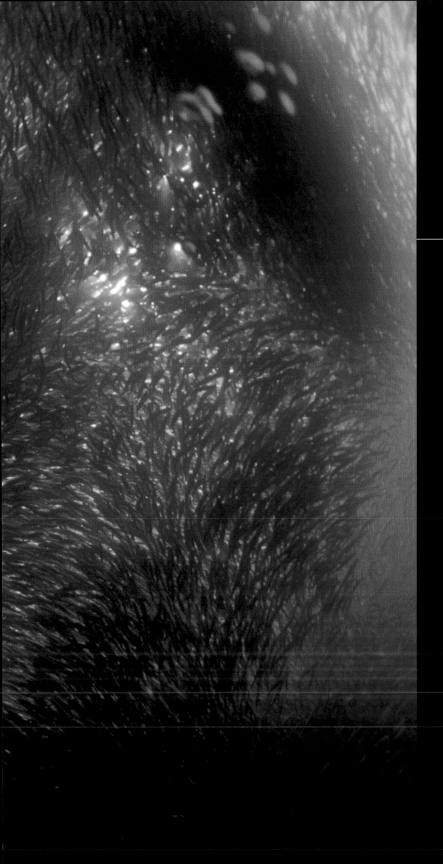

SAFETY IN NUMBERS

When predators come upon a swarm of small fish, which seem to be following some invisible leader, the result is usually a blood bath. Still, even the largest and hungriest predator can only capture a small portion of the huge swarm, so clearly there is strength in numbers.

FAMILY TIES

For a fish laying millions of eggs, it is impossible to make sure that all of its offspring survive. The best that can happen is that two or three survive long enough to procreate again. It is quite different with animals that only give birth to a few young. Parental care is far more common in such circumstances, often in the form of group participation. That is exactly what happens in the arctic when a wolf pack tries to attack a herd of musk oxen. The adults encircle their young, face outward and form a virtually

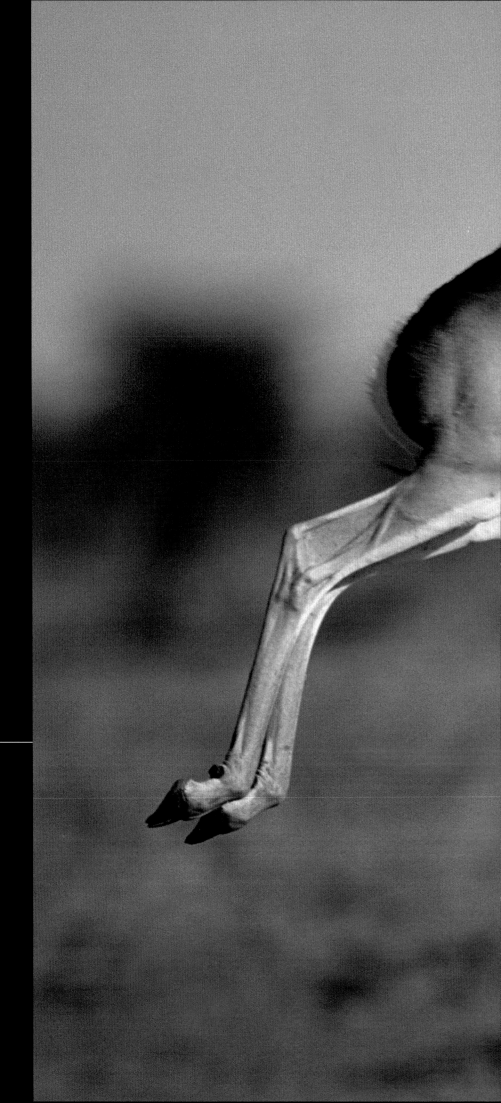

LEAP INTO FREEDOM

It's hard to believe but when chased by a predator, the African antelope can leap up to 13 feet (4 meters) in the air. Unfortunately, this behavior also makes the antelope far more visible to hunters by drawing attention to itself. Normally the antelope is not easily spotted in the tall grass of the African savanna. But perhaps the leaping antelope sends another message to its predators: "I know you are there, but I don't care since I can outrun you, so maybe you should reconsider chasing me."

MY BARK IS WORSE THAN MY BITE

To fend off unwanted encounters, its sometimes enough to just be noisy. Despite the old saying, "barking dogs don't bite," a loud noise does draw attention and perhaps some respect, especially if you can convince your foe that you are larger and stronger than you really are. This simple, very old tactic is practiced by the Australian frill-necked lizard. Though barely one yard in length, the lizard opens its large, ruffled neck folds and makes hissing noises when an intruder approaches, giving the appearance of a terrifying dragon.

WHO WILL BE THE VICTOR?

It is often assumed erroneously that the Indian mongoose is immune to the bite of poisonous snakes. Should the snake succeed in biting the mongoose, it would succumb to the venom just like people or dogs. But with lightning speed, the mongoose bites the snake in the back of the neck to prevent it from striking. Usually the mongoose is the victor in such dramatic encounters. As if to savor its victory as long as possible, it stays on the scene until it has fully devoured its foe.

ESCALATION

Thanks to their protective shells, sea snails have withstood attacks by crabs for the past two hundred million years. Although in the past snail shells were less robust than they are now, so too were the pincers of their crab adversaries. In parallel with the progressive hardening of snail shells over time, crab pincers have also evolved into stronger weapons. That has made little difference though, for in addition to thickening, snail shells have become smoother and much more difficult to grasp. Because of this, cowrie snails only have to retreat into their hard shells to avoid being harmed by crabs.

A PRICKLY SITUATION

Thanks to its prickly exterior, the sea urchin is well protected, an essential survival feature for a creature that can neither hide nor flee. Along similar lines, the tropical hedgehog has even more effective defenses, thanks to its long, brittle spines, which can easily bury themselves in the skin of an aggressor, release poison and cause painful wounds.

HARD SHELL, SOFT CORE

Strong armor to protect its back and other exposed body parts is probably one of the simplest animal defenses. But this comes at a cost, because such armor inhibits quick escapes and is not suitable for a predatory life style. For armadillos this has led to a specialized existence, mostly nocturnal and underground, where it can readily crawl away, thanks to its strong claws. It feeds primarily on insects, and is distantly related to the anteater.

DOWN IN THE DEEP

Many sea worms spend their entire lives in tubes from which they only extend a small distance. Attracting plankton through their colorful crowns, these worms feed on plankton through a mouth opening located on their underside. However, this is not an assured food source. Tube worms live at much greater depths, in oxygen-poor waters near hot ocean vents, from which molten rock rises. This is a very deadly environment for most animals, yet ideal for certain types of bacteria. Giant tube worms have adapted to this extreme environment. They lack mouths and digestive systems and survive on symbiotic bacteria colonies housed in their bodies. The worms nourish themselves on the products of these bacteria.

WHEN WE ASK OURSELVES WHETHER OTHER LIFE FORMS EXIST IN OUR VAST UNIVERSE, WE ARE REMINDED THAT LIFE AS WE KNOW IT HAS EVOLVED TO FIT THE ENVIRONMENTAL CONDITIONS THAT PREVAIL ON OUR PLANET TODAY. THIS MUCH IS CERTAIN: OVER THE SPAN OF MORE THAN THREE BILLION YEARS OF EVOLUTIONARY HISTORY, EVERY LIFE FORM THAT WAS LESS FIT THAN OTHERS WAS REPLACED AND EVENTUALLY BECAME EXTINCT. WE CANNOT ENVISAGE ANY LIFE FORM THAT COULD SURVIVE UNDER CONDITIONS THAT EXIST ON THE MOON TODAY, OR ON ANY OTHER PLANET IN OUR SOLAR SYSTEM. THIS IS LARGELY DETERMINED BY THE NATURE OF THE MOLECULES AT THE BASIS OF TERRESTRIAL LIFE. MOST PROTEINS BECOME DENATURED (THAT IS TO SAY, THEY LOSE THEIR FUNCTIONAL STABILITY), AT TEMPERATURES JUST A FEW TENTHS OF A DEGREE ABOVE BODY TEMPERATURE. SIMILARLY, THE COILED STRANDS OF DNA SEPARATE UNDER ELEVATED TEMPERATURES.

In view of the high water content that all organisms exhibit (two-thirds of body weight in humans and up to 98 percent in jellyfish), it's hard to imagine what would happen to an organism when the temperature drops below the freezing point of water, or when it rises above the boiling point. Although cells are not filled with distilled water, nonetheless they freeze at just a few degrees below zero Celsius. This results in ice crystal formation and irreversible damage within the cell. The situation is equally bad when temperatures approach the boiling point, and indeed most organisms do not survive temperatures well below that point.

There is also an optimal temperature range wherein cellular metabolism is fastest and most efficient. That is because the efficiency of enzymes, those sensitive cellular molecules, is at its highest at those temperatures, a factor crucial to the overall well being of organisms.

Mammals and birds have evolved a very interesting mechanism to mitigate against temperature extremes that are too high or too low. They have developed internal temperature regulatory systems that allow them, with only small fluctuations, to maintain their body temperatures within the optimal range for peak enzymatic efficiency. By these mechanisms, warm-blooded or homoeothermic animals are relatively independent of environmental temperature fluctuations.

As expected, temperature regulation of this type, shared by humans and other mammals as well as birds, requires significant energy expenditure, particularly when the external temperature drops significantly below body temperature. This can also have some damaging consequences, by providing homoeothermic animals a distinct advantage over animals that don't have this ability, and whose activity is more dependent on external temperatures.

Picture a bird on the hunt for a lizard. When the sun warms the rocks on a nice summer day, things are relatively even, since the lizard can scamper as quickly into the underbrush as the bird can dive from on high. However, on cooler occasions, perhaps in the early morning, or on a cloudy day, the bird can strike just as rapidly as on a sunny day, but the cold-blooded lizard moves more slowly and becomes easy prey.

Mammals and birds can also remain active when temperatures drop, provided they find sufficient food to maintain their elevated body temperatures. As resources diminish seasonally in forests or grasslands, squirrels and marmots retreat to their winter hibernation sites, deer descend to lower elevations and birds ready themselves for migration. Other species, who spend the entire winter at high elevations despite snow cover and ice-cold temperatures, are in constant search for food. Some, like ermines and snowshoe hares, have evolved special winter coats to avoid being seen by predators.

When it comes to even hardier environments, like polar ice regions, no cold-blooded animals can survive. This is the domain of polar foxes, polar bears and penguins. Enemies are rare in these environments and resources even rarer. There is no vegetation and there are few land animals. Although there are comparatively few species in arctic environments, none could survive without outside resources, namely the sea. Penguins find fish and squid there, while polar bears wait by holes in the ice for seals to surface.

Denizens of other extreme environments are also dependent on external resources for survival. After searching for food all night in the outside world, bats seek seclusion in caves, where small invertebrates in turn feed on bat excrement and carcasses that have fallen to the cave's floor.

Challenging conditions also prevail deep in the ocean, where the sun's rays do not reach. Because algal photosynthesis is impossible, nourishment must come from above. In part this comes from fish and other animals that move up and down in the water column, but mostly it comes from the remains of surface organisms that sink to the sea bottom and are consumed by the deep dwelling fauna.

One thing that cave dwellers and deep ocean fauna have in common is lack of sunlight. Both are inhospitable environments, where survival is difficult but not impossible. Cave dwellers generally have poor vision or are totally blind, like the cave salamander. The same holds true for crustaceans that inhabit underground waters and even the rare fish species that have adapted to these extreme conditions. Invertebrates that live underground, such as centipedes and millipedes, are also blind. Though lacking eyes, they move along briskly, thanks to their comparatively long legs and above all, the long feelers that help them assess the size and distance of objects in their path.

In addition to making do with limited food resources, cave dwellers have also adapted to their environment through significantly shortened development times from egg to adult form. Cave salamanders, for example, do not undergo metamorphosis. In contrast to its surface-dwelling counterpart, which begins life as a tadpole with gills and then transitions to a land animal with lungs, the cave salamander retains its gills as an adult and lives only in water.

Cave-dwelling animals can survive their harsh environment thanks to specialized adaptations that allow them to live with limited resources. The question, of course, is what drove their surface-dwelling ancestors to the subterranean environment? For some it was probably triggered by the drastic cooling of the earth over the last million years, as indicated in the geologic record, which led to several successive periods of glaciation. The underground caves probably served as refuges for small land animals.

There are two other possibilities concerning the ancestry of cave dwelling animals. One is that changing environmental temperatures and humidity made living underground more conducive to their survival. The other is that competition from other species drove them there. Either way, it is evident that even very harsh environments can offer advantages to their inhabitants.

This also applies to the dark depths of the ocean, where at first glance environmental conditions appear less than ideal for many species. However, deep sea conditions remain fairly constant over long periods of time, which makes it less risky for species to evolve and adapt to them. One difficulty is finding mating partners. In deep waters, the search for mates in near total darkness is challenging at best. Yet even here, where the sun never shines, dim lights illuminate the deep. This light is given off by the animals themselves, be they shrimp, squid or fish. Many fish in particular are covered with bioluminescent spots on their bodies or have specialized light-emitting organs over their mouths. In addition, they possess huge eyes that are sensitive to dim light sources carried over long distances.

When two partners find each other under these circumstances, it is very important that they pair up, since the opportunity might never present itself again. It would be preferable if they paired for life. This is indeed what happens with certain fish that are related to the monkfish. As soon as the much smaller male encounters a female, it attaches itself permanently to her body. The attachment is tissue-linked, making the male literally an appendage of the female's body. This amazing linkage is only possible because there is no effective tissue rejection among these fish, as is common with organ transplants in other animals.

Over the past 30 years, deep sea research has discovered a world of organisms living in extreme environments. This is the result of hydrothermal vents deep on the ocean floor, where hot, sulfur-rich water is present in the total absence of oxygen. One would expect bacteria to thrive in this environment and derive energy through chemical reactions with hydrogen sulfide. That is exactly what happens and there is indeed an abundance of bacteria living there. What was not expected was the complex ecosystem of soft-bodied animals and tube worms that also thrive in these deep sea environments. All these creatures depend entirely on the bacteria as their only food source, given that they are the only organic material in this extraordinary environment. The bacteria are not eaten and digested, however, but are incorporated directly into the bodies of the tube worms in a symbiotic union. The bacteria thus serve but one function: to permit their hosts to survive on the ocean floor.

Thus we have here an alien ecosystem that survives in total darkness, under unfavorable temperatures, in the absence of oxygen and with a near-total lack of nutrients. It is clear that the absence of water would be far more threatening for the survival of animal and plant life than these adverse conditions.

Even when water appears abundant, however, it may not be available to organisms. This is particularly evident in shallow ponds when water becomes excessively salty as a result of evaporation. Since cells normally dehydrate under such conditions, only organisms with cells capable of storing or replacing the requisite water can survive. Only a few species can do that.

In other environments, water can totally disappear for some time. Life is hard enough in swamps, but at certain times of the year they can dry out completely. Amphibians like toads generally cope best under these conditions, since they really need liquid water only as long as it takes for their eggs to be laid and the tadpoles to mature. As adults they do just fine on dry earth. Some species, like African lungfish, have adapted to live in mud and slime by using lungs in addition to gills.

Finally there are arid desert environments with endless expanses of sand or stones where it never rains. Every survival strategy imaginable to conserve water can be found here. Predators like snakes and scorpions capture prey with potent poisons. They can live off the small amounts of water obtained from the bodies of their captured prey.

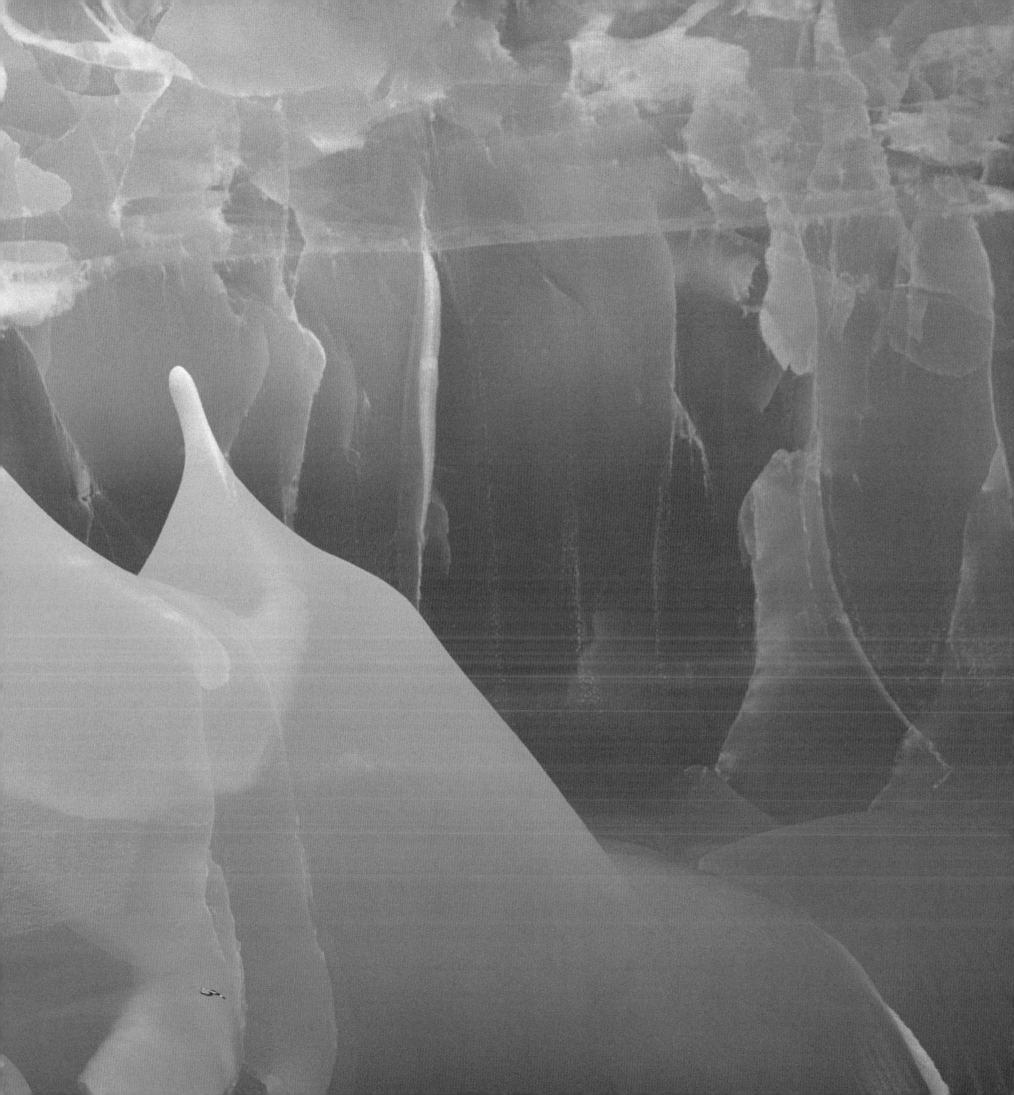

THE TIP OF THE ICEBERG

Incredible physical feats are routine for penguins in Antarctica. Male emperor penguins stand upright for months without nourishment to protect and incubate their single eggs in frigid temperatures. Penguins are also extraordinary divers, who can stay underwater for 15 minutes without breathing, long enough to seek food as deep as 650 feet (200 meters) below the surface.

ALWAYS ON THE MOVE

Life can be very strenuous for small birds, especially for tiny hummingbirds. Their wings are in constant motion. They must hover from flower to flower, because the delicate blooms would buckle under the weight of the humming bird, even though they are so light. Thanks to its very long beak and its long, split tongue, the bird gathers nectar and small insects that reside in the flower. With heart rates in excess of 1,200 beats per minute, it is little wonder that hummingbird bodies must cool down at night and rest until the following day.

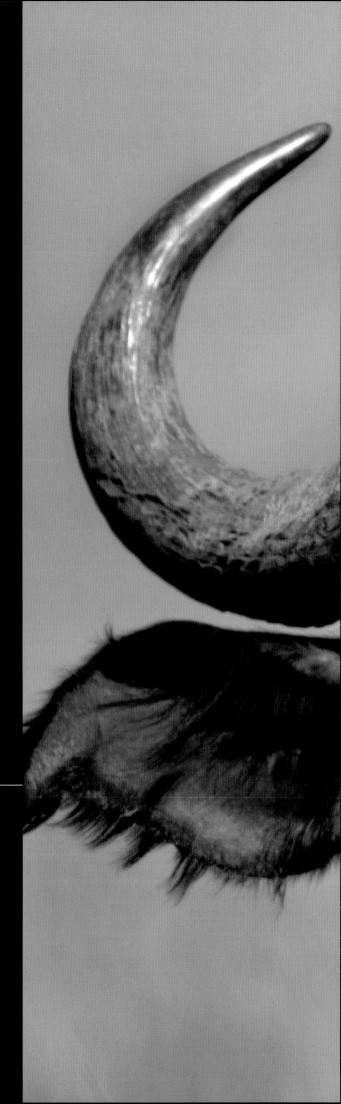

IN SEARCH OF NEW SHORES

Survival for many animals means they have to pull up camp from time to time and move to distant locations. Animals that travel seasonally select their new location with purpose. Over thousands of years, migratory birds have flown their routes so often that they now do it instinctively. They know precisely which routes to take, and they use the sun, stars and the earth's magnetic field for orientation if they get lost or need to stop to rest. They also know instinctively when and where to land, and how long they must stay before the seasons change and it is time to return home.

FRIENDS IN NEED

Sometimes the most unlikely partnerships are established between animals that could not be more different. They coexist side by side, even though one could easily kill the other. We find one example of such friendly coexistence in the African savanna. The yellow-beaked oxpicker, a relative of the starling, spends its entire life on the backs of such giants as water buffalo or rhinoceros, feeding on parasites in the animals' fur. Another African bird, the crocodile bird, even dares to explore the mouths of crocodiles in the Nile in search of food.

SOARING TO GREAT HEIGHTS

The condor manages two impressive feats. First, it supports a wingspan 15 times larger than that of a chickadee and a thousand times heavier. Second, it flies at such dizzying altitudes that breathing seems impossible and yet the condor proves us wrong.

TRACKS IN THE SAND

It is not easy to move over sand without sinking. That is why the screw horn antelope has extremely wide hoofs so as to spread its weight over a larger area. That is also the reason why the sand skink, a strong, short-legged lizard, dives into the sand and virtually swims in it like a fish. Moving even more lightly, the sidewinder rattlesnake puts fear into the hearts of anyone who spots its signature zigzag tracks in the sand.

AND WHEN IT'S TOO HOT....

Cold is tolerable either when body activity is reduced to a minimum or if internal warmth is available. But how do animals cope with extreme heat in the desert? Often it is not enough to stay out of the sun or defer activity to the evening. Even in the shade with a minimum amount of activity, body temperatures can climb far too quickly. The only solution is to find a way to release body heat. That is the prime function of the huge ears of the cape and desert foxes, two small desert-dwelling species.

LIVING BETWEEN THORNS

Where to nest, when snakes are a threat and there are no trees to offer protection? In arid terrain where plants are sparse and covered with spines, birds have few options. To avoid snakes on the ground, they prefer to nest among the spines of the giant saguaro cacti.

TWO LEAVES IN A HUNDRED YEARS

In light of the extreme water shortage in many desert ecosystems, many plants have abandoned regular leaves in favor of spikes and thorns. Chlorophyll-driven photosynthesis must take place in the plant stems, providing the plant has a stem that stands above ground and lacks branches. This is definitely not the case with welwitchia, an unusual denizen of southern West African deserts. Its stem barely rises three-quarters of an inch (2 centimeters) above the arid terrain. The plant only has only a few pairs of very elongated leaves, which grow slowly but constantly. In as harsh an environment as the desert, it would be highly inefficient to shed leaves and grow entirely new ones.

HEAVYWEIGHT

When it comes to size, there are limits for every type of animal, be it mammal, bird, fish or butterfly. Whether maximal or minimal, these sizes ranges cannot be exceeded. Among land mammals, this ranges from the miniature shrew, which, including its tail, measures barely 2 inches (5 centimeters), to the male African elephant, which can reach a height of 6½ feet (2 meters) and weigh up to 5 tons. Four legs could not carry much more mass than that. A totally different situation prevails in the sea, where the blue whale holds the record, measuring 108 feet (33 meters) in length and weighing 190 tons. More than that is probably not possible.

I CAN'T SEE ANYMORE

It is not only the total absence of light and the diurnal cycle that make life in caves extremely difficult. More serious is the scarcity of nutrients, particularly in the deepest caves. Even bat guano and cadavers are absent here as possible food sources. Such conditions dictate the utmost frugality. That is why deep cave dwelling species lack even eyes, for which they have no use in the perpetual darkness.

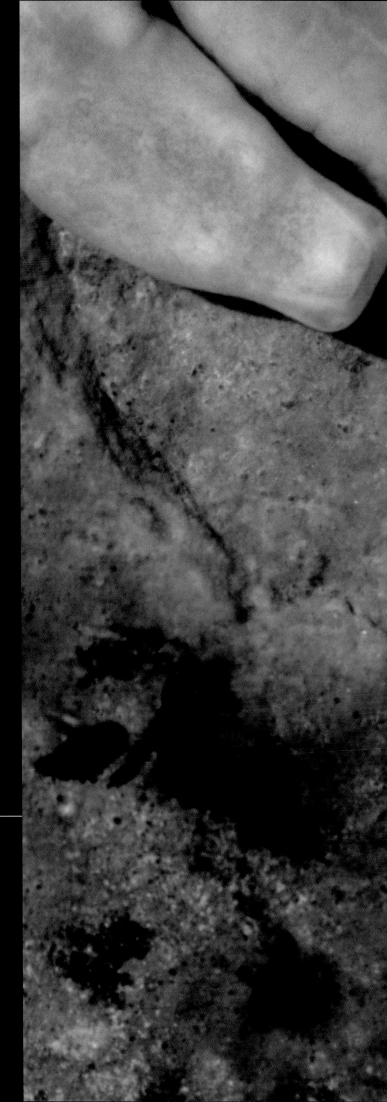

I SEE THE LIGHT

Not unlike the situation in deep grottos, the ocean depths also lack both light and nutrients. Yet scattered sources of light do occur here, emitted by certain crustaceans and jellyfish, but primarily by species like the angler fish. This fish is equipped with a bioluminescent organ on its upper jaw, ideal for attracting other fish and crustaceans as prey.

CHANGE

ANIMALS AND PLANTS IN MODERATE CLIMATE ZONES FACE NEW CHALLENGES, DIFFICULTIES AND OPPORTUNITIES EACH SEASON. IN AUTUMN, LEAVES CHANGE COLOR AND FALL, AND SUNLIGHT CAN REACH MANY MORE SHRUBS AND UNDERBRUSH PLANTS THAN BEFORE. THEY MATURE AND PROVIDE NOURISHMENT FOR COUNTLESS SMALL MAMMALS AND BIRDS. AT THE SAME TIME, THE ABSENCE OF FOLIAGE MAKES IT EASIER FOR PREDATORS TO SPOT THEIR PREY. IN WINTER, TEMPERATURES DROP AND RESOURCES BECOME SCARCER. SPRING BRINGS MODERATE WEATHER AND MORE FOOD RESOURCES, BUT IT ALSO BRINGS MORE COMPETITION AS ANIMALS LOOK FOR MATING PARTNERS. THEN SUMMER ARRIVES, A SEASON THAT CAN BE TOO SHORT AT HIGHER LATITUDES OR IN POLAR REGIONS. YET, IT MAY SEEM TOO LONG FOR SPECIES THAT GRAZE IN HOT AND ARID REGIONS ON LITTLE MORE THAN PARCHED VEGETATION, WHERE IT IS HARD TO HIDE FROM THE SUN OR ENEMIES. IN SUCH REGIONS, NOT MANY SPECIES CAN REMAIN ACTIVE YEAR ROUND. OFTEN THEY MUST MIGRATE TO FIND MORE CONDUCIVE LIVING CONDITIONS. HE BEST CHANCE OF SURVIVAL UNDER SUCH EXTREME CONDITIONS, BE IT DURING THE COLDEST OR THE HOTTEST TIME OF YEAR, IS EITHER TO HIBERNATE OR TO REDUCE ACTIVITY TO A MINIMUM.

RETURN TO THE SEA

Some islands are hundreds of kilometers from the mainland. Only birds that can live off the sea can undertake such a long voyage. They must fish for food and be able to take their rest by floating, without any access to dry land. During mating season these birds gather by the thousands on tiny spits of land like Ascension Island in the mid-Atlantic Ocean. They remain there only as long as needed to raise their young. As soon as they are old enough the birds go out to sea and only come back to land during mating season.

Similar strategies apply to many plant species that only have a few months, sometimes only a few weeks, in which to grow leaves, elongate stems, bloom and bear fruit. They die after producing and scattering their seeds. Other species let their leaves, flowers and stems die but keep bulbs or tubers underground to propagate again the following spring. Many butterfly species have adopted similar strategies. They over-winter either encased as eggs or as pupae, to emerge as full adults with wings during the warmer season.

It is considerably more difficult for warm-blooded animals to enter a state of diminished activity, with notable exceptions. Bears, marmots, squirrels and mice are examples of mammals that can scale back vital activities like breathing and circulation to the minimal levels needed for survival.

Because their physiological activity is far higher than that of mammals, birds cannot hibernate, but they have evolved alternate means to survive winter. They can fly over great distances and migrate to warmer climate zones, and then return to the temperate climates to reproduce during the spring.

All plants and animals have adapted to withstand seasonal changes. For those of us living in temperate climates, dramatic temperature changes are the most obvious reminders of the time of year. In latitudes closer to the equator, seasonal changes are usually manifest through heavy rains or dry months.

The consequences of this can be quite dramatic. In the Amazon region, for example, the extreme rains can cause extensive flooding of the river and its tributaries. Trees and vegetation on the riverbanks are sometimes under 16 feet (5 meters) of water. What sorts of adaptations are needed under those circumstances?

These periodic flooding episodes are no less stressful than the arrival of ice and snow in the forests and mountain meadows. Few animals are as adaptable as amphibians, who can transition from water to land with ease. Animals really only have two survival options: spend the harsh season in hibernation or move away. This does not necessarily imply migrating far away. Some species can climb to the top of trees in advance of the flooding and survive in the highest branches until the waters subside again.

The borders between land and water are very challenging to many species as they pass through different life stages. For example, mosquito larvae must have water to mature, while the adults spend most of their lives on land. The same holds true for dragonflies, frogs and various species of salamanders.

When mosquitoes emerge from the water after pupating, they are in a very precarious situation. Many unforeseen predators lurk in the deep. Equally dramatic are the quick adaptations the adult mosquito now faces: moving in air rather than water and using trachea and lungs, rather than gills, to draw in oxygen. These transitions do not take place in a calm setting, but in an environment where the adult insect must adjust very quickly to entirely new circumstances. This is reminiscent of how migratory birds must rapidly adjust to change. They must undertake their journey well before the first snowfall or, due to diminished food sources, risk becoming too weak to travel to warmer climes.

In order to survive, it may not be enough to migrate, hibernate, metamorphose, or simply to change location. It is also crucial to anticipate change. That is why migratory birds depart when nutrients are still plentiful, having stored enough fat reserves to permit them to travel long distances over oceans and deserts before reaching fresh food sources. Likewise immature frogs and salamanders practice breathing through their new lungs or skin well before losing their tadpole gills. Mammals living in temperate and cold climates prepare for change before the seasons shift. Mating seasons are timed in such a way to insure that the offspring are born at the best possible times for survival.

The changing seasons are not only manifested through changes in temperature and day length, but also through internal physiological adjustments. These are not rapid and reversible changes in color or shading, such as a chameleon or flounder might undergo to blend in with its surroundings. In mammals, gradual changes include hormonal adjustments as the animals mature into adults. In insects, several larval stages lead to adulthood, as well as the caterpillar's spectacular metamorphosis into a butterfly. These are all one time, irreversible changes.

Freshwater eels are spawned in the Sargasso Sea in the Atlantic Ocean and migrate to the European coastline and the calm waters and abundant vegetation of brackish ponds or shallow river waters. Later they leave the rivers where they have spent most of their lives and migrate back to the sea to reproduce. During its first long migration, the eel does not initially resemble the long "snake" we are familiar with, but has a flat body, not unlike a leaf. Not surprisingly, 19th century scientists thought they were two different species of fish.

Some species undergo unexpected changes, like the fly and the sea urchin. These creatures undergo major morphological or lifestyle changes, and as larvae they also occupy or destroy the bodies of much larger animals.

In the case of flies, almost all structures present in larvae are destroyed during the sessile pupa stage, including the skin, muscles and digestive organs. The nervous system remains intact, but the brain goes through a restructuring process, because as an adult it will have to process information through its anterior feelers and large eyes, which was not necessary in the larva stage. The rest of the body is more or less newly formed, not only the eyes and feelers, but also the legs, wings and reproductive organs. This is followed later by a new gut, muscles and an entirely new exterior. The larva remains are not completely lost: though immobile, like the wrappings of an ancient mummy, the pupa nonetheless supplies essential components for growth as the products from its molecular breakdown are utilized in the formation of the new adult body.

No less spectacular is the metamorphosis of the sea urchin larva, whose appearance is remarkably different from that of the adult. Delicate and transparent, with a two-part symmetric body, the larva nonetheless contains all the information needed to form the adult's five-sided symmetric body. The larva initially lives on plankton. Eventually the larva settles to the sea floor and the outer parts are destroyed, while the tiny snail inside matures and is freed to begin life independently.

Amid all their complexity, these changes share a common aspect: they are irreversible. A fly pupa can only develop into an adult and not revert to a larva. Similarly, the mature sea urchin cannot shed its heavy, calciferous exoskeleton and return to life as a larva.

There are species where development appears not to be irreversible, species which appear to remain eternally young. Jellyfish fit this category. They are the spawn of an apparently different life form, one that is anchored to the sea floor, while the jellyfish itself can move freely through the water. The juvenile, sessile form of this animal, resembles thousands of tiny flower-like corals. This polyp is a tiny creature with a ring of tentacles and an oral cavity. Normally, jellyfish originate from polyps and not the other way around, but there are exceptions to this rule. Under the right conditions some species of jellyfish can revert to the polyp stage, making it difficult to determine which is juvenile and which is the adult form. So the notion that change is irreversible, which applies to humans and other vertebrates, does not always apply to other animal types.

Though probably rare, life on earth has had to adapt to catastrophic changes on a planet-wide basis. That was the case about 65 million years ago, when dinosaurs and countless other species perished when the earth was struck by an asteroid.

That was not the first time mass extinctions have taken place on earth. Paleontologists have identified at least five such occurrences. The one near the end of the Cretaceous period caused the extinction of al least 60 percent of all animal forms, including the dinosaurs. At that point many of the plant and animal species of prior geologic eras were eradicated, making way for the fauna and flora that have evolved since. But this was not the earliest nor the most severe occurrence of this nature. The most drastic change in the biosphere was the massive extinction at the end of the Paleozoic period, 250 million years ago. Not even 10 percent of existing animal species survived, including both marine and land forms. Still, life did not perish completely. Similarly, the earth managed to return to health once more after the most recent ice age.

Clearly the greatest changes are taking place right now, a time where man plays the leading role. In many regions on earth massive changes in the environment are either caused, or at least mitigated, by human actions. Yet not everything that humans have touched has turned to desert. Human activity has generated new habitats, some perhaps unstable or splintered, but not necessarily adverse either. Habitats created by humans are not just havens for caterpillars or grasshoppers, but also many birds, which do better in city parks and gardens than many other places. Such habitats have changed the behavior and life cycle of many birds. Life is incredibly adaptable and can survive even major environmental changes.

A CHANGE OF CLOTHES

If a chameleon were human, the question would come up as to how it managed to change its appearance and identity so quickly and frequently in the course of its life. Actually, this unusual lizard really does nothing more than change its clothes, something far less complicated relative to what happens when a tadpole changes into a frog or a caterpillar into a butterfly. Will they even recall what they were like once they become adults?

A LIZARD WITH A PAST

The giant marine iguanas on the Galapagos Islands look very much like ancient reptiles. Had Charles Darwin not stopped on the shores of this unknown archipelago during his tour around the world, the history of this lizard and other remarkable creatures of these islands might not have come to light. Their ancestors were brought across the sea to the islands under very unusual circumstances. Once there, the iguanas had to adapt to the sparse island environment and survive on a diet of algae and seaweed. Each day they wait for the moment when low tide exposes these salty morsels.

A VARIED MENU

Bitter, unattractive or downright poisonous, all of these are defense mechanisms adopted by many leafy plants to discourage being eaten, whether by
snails, rabbits or caterpillars. And yet there are animals that can feed on plants
that others would avoid, although there can be limits here as well. Koalas, the
Australian marsupial bears, can only stomach eucalyptus leaves for a short
time each year. Fortunately there are hundreds of different eucalyptus species
whose leaves are palatable at different times of the year, forcing koalas to
migrate to different places in order to graze on edible vegetation.

A BUILT-IN INCUBATOR

In comparison to other mammals, a newborn kangaroo is clearly quite immature and has all the characteristics of a premature birth. In fact, it could not survive were it exposed immediately to its environment. Yet nature has supplied kangaroos and related species with a built-in incubator, a pouch wherein the newborn can suckle milk, rest and mature until it is ready to live on its own.

ENERGY-SAVING MEASURES

For some mammals living in temperate climates, the onset of the cold season signals the interruption of activities. Now is the time to conserve energy: find a suitable location to hibernate and reduce one's body temperature. It would simply not be energy-efficient to maintain regular body temperatures in the customary 95 to 104° F (35 to 40°C) range. Animals like this mouse lay on adequate fat reserves before the start of winter. In olden times, fattened field mice were considered as a real delicacy at the onset of winter.

THERE'S NO PLACE LIKE HOME

Many fish experience remarkable changes during their lives. Take the fresh-water eel. It begins life in the Sargasso Sea, migrates across the Atlantic Ocean and the Mediterranean Sea, and ends up in European rivers, where it spends several years. Upon reaching adulthood, it resumes its migration back to the sea. Salmon undertake exactly the opposite journey. Although shorter in distance, the trip is just as adventuresome. When returning to its birth site, it must struggle upstream against the current in order to finally lay its eggs in the calm waters of the river's source.

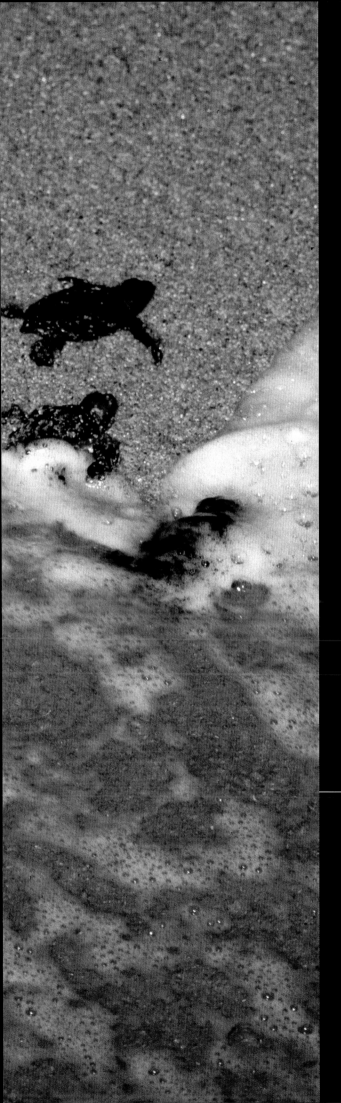

LIFE IS AN ADVENTURE

Some of the most spectacular occurrences over the millions of years of evolution are the return to land of animals previously living in aquatic environments. Likewise many land animals have returned to the water, often with great success. Examples of the latter include whales, dolphins and giant sea tortoises. The adaptation to life in the water has not always been complete. The sea tortoises return to land to lay their eggs but, upon hatching, the young tortoises find their way back to the sea.

NEITHER FISH NOR FOWL

During the Devonian period, when the first fish attempted to transition to land, they probably developed a primitive lung in addition to their gills. They would be similar to the six species of freshwater lungfish found today in South America, Africa and Australia. A primitive lungfish was probably the ancestor of land vertebrates. Later on other adaptations evolved, like the flying fish whose modified fins serve as wings, or the small gobies with their large protruding eyes, typical of many mangrove species. Mangrove life includes many borderline examples, with fish leaving the water to gasp for air, and trees growing in the salty waters with greatly extended roots.

METAMORPHOSIS

Not much remains unchanged within the seemingly quiescent pupa during the development of an adult. The muscles of the larva are transformed into the new muscles that will be used by the butterfly for flying. Similarly the larva gut is transformed from one that digested leaves to nourish the caterpillar, to one that will process flower nectar to feed the butterfly. So, is the caterpillar and butterfly one animal or actually two?

BETWEEN TWO WORLDS

Every metamorphosis represents a transition and a link between two different forms of existence, which in hindsight often seem diametrically opposed. Few metamorphoses are more spectacular, however, than when a water-dwelling larva transforms itself into a delicate insect with wings. There is a stage in which the insect can no longer survive in the water but does not yet have wings to take flight. At that point the insect resides between two worlds, and belongs to neither.

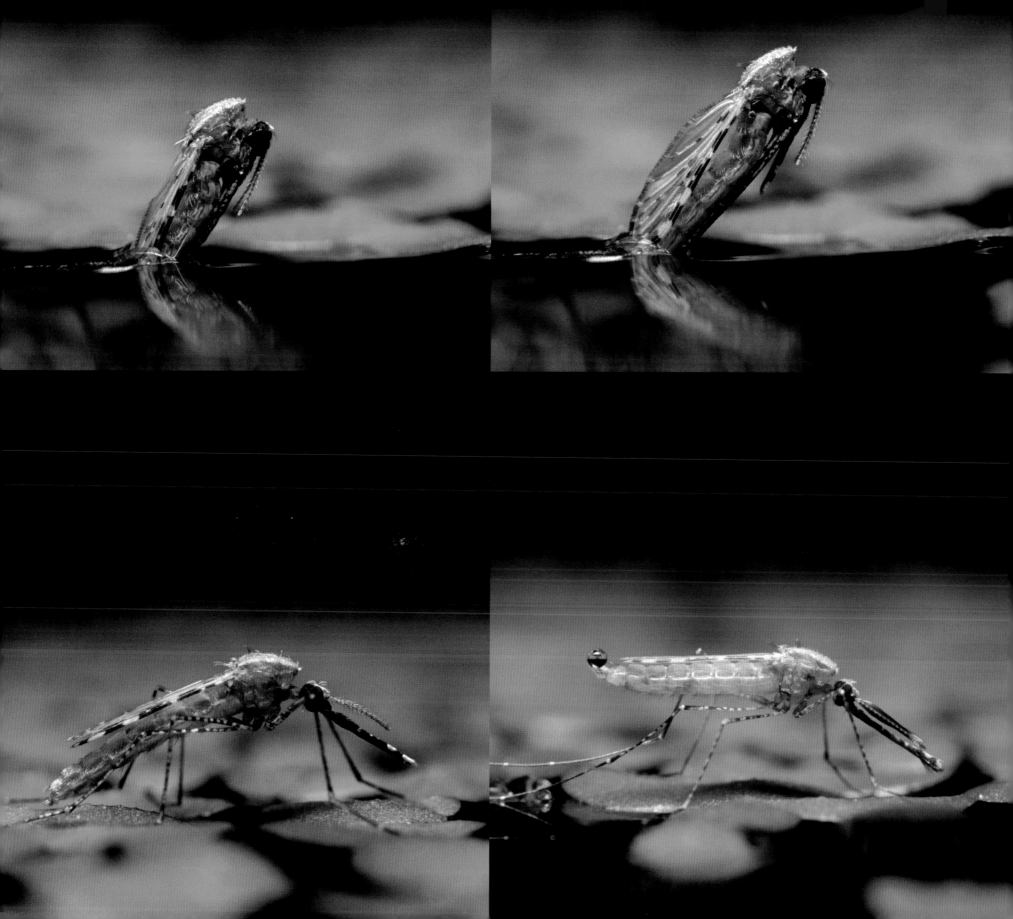

AT HOME IN THE CITY

Step by step the presence of humans has changed the surface of the earth. Among the many changes are the abrupt appearance of concrete deserts and large tracks of homogenously cultivated land, as well as pollution of soil, water and air. For many species these changes have been far too large and rapid. Other species have adapted and found new food sources or an escape from their competitors. In cities, rats, blackbirds and seagulls have found life easier and safer than many people.

PHOTO INFORMATION

AND CREDITS

– 4 –
Portuguese Man of War
Physalia physalis
Hydrozoa: sinophore physliidae
Tropical and subtropical waters,
Mediterranean
© NHPA/PHOTOSHOT, A.N.T. PHOTO LIBRARY

– 8 –
Archerfish
Toxotes jaculatrix
True bony fishes: perch-like toxotidae
South and Southeast Asia, Philippines,
Indonesian archipelago, Australia
© NHPA/PHOTOSHOT, STEPHEN DALTON

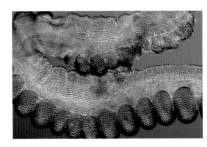

14 – 15
Portuguese Man of War
Physalia physalis
Hydrozoa: Sinophore physliidae
Tropical and subtropical waters,
Mediterranean
© NHPA/PHOTOSHOT, IMAGE QUEST 3-D

– 16 –
Golden-ringed Dragonfly
Cordulegaster boltonii
Insect: Dragonfly Cordulegasteridae
Europe, Asia
© NHPA/PHOTOSHOT, LAURIE CAMPBELL

– 17 –
Golden-ringed Dragonfly
Cordulegaster boltonii
Insect: Dragonfly Cordulegasteridae
Europe, Asia
© NHPA/PHOTOSHOT, JORDI BAS CASAS

18 – 19
Crab spider
Misumenops sp.
Spider: Web-spinning Thomisidae
Worldwide
© NHPA/PHOTOSHOT, JAMES CARMICHAEL JR.

20 – 21
Peacock mantis shrimp
Odontodactylus scyllarus
Crustacean: Peacock mantis shrimp
Odontodactylidae
Indo-Pacific Ocean
© NHPA/PHOTOSHOT, PETE ATKINSON

22 – 23
Orca
Orcinus orca
Mammals: Whale Delphinidae
Antarctic waters
© NHPA/PHOTOSHOT, MARK CARWARDINE

24 – 25
Northern Pike
Exos lucius
Bony fishes: Pikes, Esocidae
Europe, Asia, North America
© NHPA/PHOTOSHOT, LUTRA

26 – 27
Jellyfish
Cyanea capillata and Aurelia sp.
Jellyfish: Lion's mane, Cyanidae
Ulmaridae
Moderate and cold oceans and seas
© NHPA/PHOTOSHOT, IMAGE QUEST 3-D

– 29 –
Green Mamba
Dendroapsis viridis
Reptiles: Arboreal Elapsidae
West Africa
© NHPA/PHOTOSHOT, ANTHONY BANNISTER

30 – 31
Spotted Gecko
Phelsuma quadriocellata
Reptiles: Squamates Geckonidae
Madagascar
© NHPA/PHOTOSHOT, KEVIN SCHAFER

32 – 33
Chameleon
Chamaeleo chamaeloen
Reptiles: Squamate Chamaleonidae
Mediterranean coastal regions,
Africa, Asia
© NHPA/PHOTOSHOT, STEPHEN DALTON

34 – 35
Tiger Python
Python molurus
Reptile: Squamates Boidea
South Asia
© NHPA/PHOTOSHOT, NICK GARBUTT

36 – 37
Mouse-Eared Bat
Myotis myotis
Mammals: Bats Vespertilionidae
Europe
© NHPA/PHOTOSHOT, STEPHEN DALTON

38 – 39
Eurasian Bullfinch
Pyrrhula pyrrhula
Birds: Finches Fringillidae
Europe and North Asia
© NHPA/PHOTOSHOT, MANFRED DANEGGER

– 40 –
Double-Crested Cormorant
Phalacrocorax auritus
Birds: Cormorant Phalacroracidae
North and Central America
© NHPA/PHOTOSHOT, HENRY AUSLOOS

42 – 43
Steller's Sea Eagle
Haliaeetus pelagicus
Birds: Birds of prey Haliaeetus
pelagicus
Northern Europe and Asia
© NHPA/PHOTOSHOT, Jordi Bas Casas

– 45 –
Common Puffin
Fratercula arctica
Birds: Puffin Alcidae
Coastal Northern Europe
© NHPA/PHOTOSHOT, Bill Coster

46 – 47
Edible Dormouse
Gliridae
Mammals: Rodents Gliridae
Europe
© NHPA/PHOTOSHOT, Ernie Janes

48 – 49
Giraffe
Giraffa calelopardalis
Mammals: Ungulates Giraffidae
Sub-Saharan Africa
© NHPA/PHOTOSHOT, ANDY ROUSE

50 – 51
Cheetah
Acinonyx jubatus
Mammals: Carnivore Felidae
Sub-Saharan Africa, Arabian Peninsula
NICK CALOYIANIS/ NATIONAL GEOGRAPHIC IMAGE

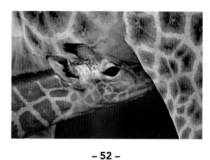

– 52 –
Giraffe
Giraffa camelpardali
Mammals: Ungulates Giraffidae
Sub-Saharan Africa
© NHPA/PHOTOSHOT, Martin Harvey

58 – 59
Blue Peacock
Pavo cristatus
Birds: Gallinaceous bird Phasianidae
Asia
© NHPA/PHOTOSHOT, JOHN SHAW

60 – 61
Stag Beetle Lucanus cervus
Insects; Beetles Lucanidae
Europe, Northern Asia
© NHPA/PHOTOSHOT, DANIEL HEUCLIN

– 62 –
Red-Eyed Tree Frog
Agalychnis callidryas
Amphibians: Frogs Hylidae
South America
© NHPA/PHOTOSHOT, KEVIN SCHAFER

64 – 65
Lion
Pantera leo
Mammals: Carnivores Felidae
Sub-Saharan Africa
© NHPA/PHOTOSHOT, JONATHAN & ANGELA
SCOTT

66 – 67
Chimpanzee
Pan troglodytes
Mammals: Primates Hominidae
Central and West Africa
© NHPA/PHOTOSHOT, MARTIN HARVEY

68 – 69
Green Sea Turtle
Chelonia mydas
Reptiles: Tortoises Cheloniidae
Oceans and sea
© NHPA/PHOTOSHOT, MICHAEL PATRICK O'NEIL

77 – 71
Pied Avocet
Recuvirostra avosetta
Birds: Wader Recurvirostridae
Europe, Asia, Africa
© NHPA/PHOTOSHOT, ROGER TIDMAN

72 – 73
Praying Mantis Mantis sp.
Insect: Predators Mantidae
Papua New Guinea
© NHPA/PHOTOSHOT, DANIEL HEUCLIN

– 74 –
Common Blue
Polommatus icarus
Insects: Butterflies Lycaenidae
Europe, Asia, North Africa
© NHPA/PHOTOSHOT, GERRY CAMBRIDGE

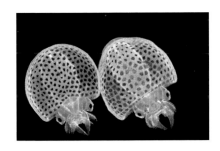

76 – 77
Diamond Squid
Thysanoteuthis rhombus
Mollusks: squids Thysanoteuthidae
Oceans
© NHPA/PHOTOSHOT, PETER PARKS

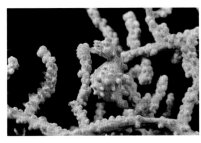

78 – 79
Pigmy Sea Horse
Hippocampus bargibanti
Chordates: Sea Horses Syngnathidae
Pacific coasts from Japan to Australia,
Indonesia
© NHPA/PHOTOSHOT, B. JONES & M. SHIMLOCK

80 – 81
Big fin Reef Squid eggs
Sepioteuthis lessoniana
Cephalopods: Squids Loliginidae
Oceans and Seas
© NHPA/PHOTOSHOT, TAKETOMO SHIRATORU

82 – 83
Emperor Penguin
Aptenodytes forsteri
Birds: Penguins Spheniscidae
Antarctica
© NHPA/PHOTOSHOT, KEVIN SCHAFER

– 85 –
Blue Tit
Parus caerulus
Birds: Passerine Paridae
Europe, North Africa, Asia
© NHPA/PHOTOSHOT, DAVE WATTS

86 – 86
Northern Elephant Seal
Miroungas angustirostris
Mammals: Carnivores Phocidae
Pacific Coast of North America
© NHPA/PHOTOSHOT, BRYAN &
CHERRY ALEXANDER

88 – 89
Cheetah
Acinonyx jubatus
Mammals: Carnivores Felidae
Sub-Saharan Africa, Arabian Peninsula
© NHPA/PHOTOSHOT, JONATHAN & ANGELA
SCOTT

99 – 91
Polar Bear
Ursus maritimus
Mammals: Carnivores Ursidae
Alaska, Siberia, Canada, Greenland,
Svalbard
© NHPA/PHOTOSHOT, ANDY ROUSE

92 – 93
Mississippi Alligator
Alligator mississippiensis
Reptiles: Crocodiles Alligatoridae
USA
© NHPA/PHOTOSHOT, THOMAS KITCHIN &
VICTORIA HURST

– 94 –
Madagascar Day Gecko
Phelsuma madagascariensis
Reptiles: Squamates Gekkonidae
Madagascar
© NHPA/PHOTOSHOT, MARTIN HARVEY

100 – 101
Sea Spiders
Parapallene sp.
Arthropods: Callipallenidae
Tropical waters
© NHPA/PHOTOSHOT, B. JONES & M. SHIMLOCK

102 – 103
Madagascar Walkingstick
Parectosoma mocqueysis
Insects: Stick insect Bacillidae
Madagascar
© NHPA/PHOTOSHOT, KEVIN SCHAFER

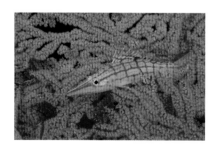

104 – 105
Longnose Hawkfish
Oxycirrhites typus
Fishes: Perciforms Cirrhitoidea
Coral reefs
© NHPA/PHOTOSHOT, MICHAEL PATRICK O'NEAL

– 107 –
Dwarf Frog
Physalaemus sp.
Amphibians: Frog Leptodactylidae
Central and South America
© NHPA/PHOTOSHOT, JANY SAUVANET

108 – 109
Hermit Crab
Paguritta vittata
Crustaceans: Decapods Paguridae
Sea of Japan
© NHPA/PHOTOSHOT, TAKETOMO SHIRATOR

110 – 111
Clownfish
Amphiprion bicinctus
Fishes: Perciformes Pomacentridae
Southeast Asia, Oceania, coral reefs
© NHPA/PHOTOSHOT, B. JONES & M. SHIMLOCK

112 – 113
Caddisfly
Phryganae sp.
Insects: Caddisflies Phryganeidae
Many regions of the Northern
Hemisphere
© NHPA/PHOTOSHOT, STEPHEN DALTON

– 114 –
Atlantic Salmon
Salmo salar
Bony fishes: Salmon Salmonidae
Western and Eastern Atlantic,
Baltic Sea
© NHPA/PHOTOSHOT, JOHN SHAW

115
Salmon Eggs
Salmo salar
Bony Fishes: Salmon Salmonidae
Western and Eastern Atlantic, Baltic Sea
© NHPA/PHOTOSHOT, GEORGE BERNARD

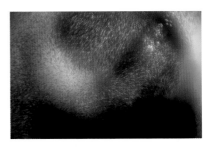

116 – 117
European Anchovy
Engraulis encrasicolus
Fishes: related to Herrings Engraulidae
Mediterranean Sea
ROY TOFT/NATIONAL GEOGRAPHIC
IMAGE COLLECTION

118 – 119
Muskoxen
Ovibos moschatus
Mammals: Ungulates Bovidae
North America, Russia, Norway
© NHPA/PHOTOSHOT, BRYAN &
CHERRY ALEXANDER

120 – 121
Springbuck
Antidorcas marsupialis
Mammals: Ungulates Bovidae
South and Southwest Africa
© NHPA/PHOTOSHOT, NIGEL J. DENNIS

– 123 –
Frill-necked Lizard
Chlamydosaurus kingi
Reptiles: Squamates Agamidae
Northern Australia, New Guinea
© NHPA/PHOTOSHOT, JAMES CARMICHAEL JR.

124 – 125
Cobra
Naja naja
Reptiles: Squamates Elapsidae
Indian Mongoose
Herpestes edwardsii
Mammals: Carnivore Herpestidae
India
© NHPA/PHOTOSHOT, DANIEL HEUCLIN

– 127 –
Black-spotted Cowrie
Cypraea nigropunctata
Mollusks: Gastropods Cypraeoidea
Coasts of Equador and Peru, Galapagos
Islands
© NHPA/PHOTOSHOT, MICHAEL PATRICK O'NEAL

128 – 129
Sea Urchins
Diadema sp.
Echinoderms: Sea Urchins Diadematidea
Karibik
© NHPA/PHOTOSHOT, B. JONES & M. SHIMLOCK

130 – 131
Nine-Banded Armadillo
Dasypus novemcinctus
Mammals: Armadillos Dasypodidae
USA to Brazil
© NHPA/PHOTOSHOT, DANIEL HEUCLIN

– 132 –
Christmas Tree Worm
Spirobranchus sp.
Polychaetes: Serpulvidae
Indo-Pacific Ocean
© WOODFALL WILD IMAGES/PHOTOSHOT

138 – 139
Chinstrap Penguin
Pygoscelis antarctica
Birds: Penguins Spheniscidae
Antarctica
© NHPA/PHOTOSHOT, BRYAN &
CHERRY ALEXANDER

140 – 141
Sparkling Violetear
Colibri coruscan
Birds: Hummingbirds Trochilidae
South America
© NHPA/PHOTOSHOT, STEPHEN DALTON

142 – 143
Canada Goose
Grus Canadensis
Birds: Goose Gruidae
North America, Canada
© NHPA/PHOTOSHOT, RICH KIRCHNER

144 – 145
Yellow-billed Oxpecker
Buphagus africanus
Birds: Passerine Strunidae
African Buffalo Syncerus caffer
Mammal: Ungulates Bovidae
Africa
© NHPA/PHOTOSHOT, DARYL BALFOUR

– 147 –
Andean Condor
Vultur gryphus
Birds: Scavenger Cathartidae
South America
© NHPA/PHOTOSHOT, KEVIN SCHAFER

148 – 149
Horned Desert Viper
Cerastes cerastes
Reptiles: Squamates Viperidae
South Africa
© NHPA/PHOTOSHOT, DANIEL HEUCLIN

150 – 151
Cape Fox
Vulpes chama
Mammals: Carnivore Canidae
South Africa
© NHPA/PHOTOSHOT, NIGEL J. DENNIS

– 153 –
Gila Woodpecker
Melanerpes uropygialis
Birds: Woodpecker Picidae
North America
© NHPA/PHOTOSHOT, DANIEL HEUCLIN

154 – 155
Welwitchia
Welwitschie mirabilis
Plant: Gymnosperm Welwitschiaceae
Southwestern Africa
© NHPA/PHOTOSHOT, DR. ECKART POTT

156 – 157
Blue Whale
Balaenoptera musculus
Mammals: Whale Balaenopteridae
Oceans worldwide
© NHPA/PHOTOSHOT, DAVID & IRENE MYERS

158 – 159
Olm Proteus anguinus
Amphibians: Protidae
Dalmatian Mountains
© NHPA/PHOTOSHOT, HELLIO & VAN INGEN

160 – 161
Black Sea Devil
Melanocetus johnsoni
Fishes: Ray-finned fishes
Malanocetidae
Tropical and subtropical seas
© SOUTHAMPTON OCEANOGRAPHY CENTER

– 162 –
Great Frigate Bird
Fregata minor
Birds: Frigate bird Fregatidae
Oceans worldwide
© NHPA/PHOTOSHOT, MARTIN HARVEY

168 – 169
Panther Chameleon
Furcifer pardalis
Reptiles: Squamates Chamaeleonidae
Madagascar
© NHPA/PHOTOSHOT,
JAMES CARMICHAEL, JR.

170 – 171
Marine Iguana
Amblyrhynchus cristatus
Reptiles: Squamates Iguanidae
Galapagos Islands
© NHPA/PHOTOSHOT, NICK GARBUTT

172 – 173
Koala
Phascolarctos cenereus
Mammals: Marsupial Phascolarctidae
Australia
© NHPA/PHOTOSHOT, DAVID HIGGS

174 – 175
Red Kangaroo
Macropus rufus
Mammals: Marsupials Macrpodidae
Australia
© NHPA/PHOTOSHOT, DAVID WATTS

– 177 –
Edible Dormouse
Glis glis
Mammals: Rodents Gliridae
Europe
© NHPA/PHOTOSHOT, ERNIE JANES

178 – 179
Chinook Salmon
Oncorhynchus tshawtscha
Fishes: Salmon Salmonidae
North American waters from Alaska to
California
© NHPA/PHOTOSHOT, THOMAS KITCHIN &
VICTORIA HURST

180 – 181
Loggerhead Sea Turtle
Caretta caretta
Reptiles: Turtles Chelonidae
Mediterranean Sea
© NHPA/PHOTOSHOT,
ANTHONY BANNISTER

– 183 –
Mudskipper
Periophthalmus barbarus
Amphibious fish: Gobidae
Atlantic Ocean
© NHPA/PHOTOSHOT, STEPHEN DALTON

– 184 – (top)
Monarch Butterfly
Danaus plexippus
Insects: Butterflies Nymphalidae
North, Central and South America, Canada,
Oceania, Europe
© NHPA/PHOTOSHOT, THOMAS KITCHIN &
VICTORIA HURST

– 184 – (bottom)
Monarch Butterfly
Danaus plexippus
Insects: Butterflies Nymphalidae
North, Central and South America, Canada,
Oceania, Europe
© NHPA/PHOTOSHOT, THOMAS KITCHIN &
VICTORIA HURST

– 185 –
Monarch Butterfly
Danaus plexippus
Insects: Butterflies Nymphalidae
North, Central and South America, Canada,
Oceania, Europe
© NHPA/PHOTOSHOT, THOMAS KITCHIN &
VICTORIA HURST

– 186 –
Mosquito
Anopheles sp.
Insects: Diptera Culicidae
Worldwide
North, Central and South America, Canada,
Oceania, Europe
© NHPA/PHOTOSHOT, GEORGE BERNARD

– 187 – (top left)
Mosquito
Anopheles sp.
Insects: Diptera Culicidae
Worldwide
North, Central and South America, Canada,
Oceania, Europe
© NHPA/PHOTOSHOT, GEORGE BERNARD

– 187 – (top right)
Mosquito
Anopheles sp.
Insects: Diptera Culicidae
Worldwide
North, Central and South America, Canada,
Oceania, Europe
© NHPA/PHOTOSHOT, GEORGE BERNARD

– 187 – (bottom left)
Mosquito
Anopheles sp.
Insects: Diptera Culicidae
Worldwide
North, Central and South America, Canada,
Oceania, Europe
© NHPA/PHOTOSHOT, GEORGE BERNARD

– 187 – (bottom right)
Mosquito
Anopheles sp.
Insects: Diptera Culicidae
Worldwide
North, Central and South America, Canada,
Oceania, Europe
© NHPA/PHOTOSHOT, GEORGE BERNARD

188 – 189
Kittiwake
Rissa tridactyla
Birds: Gulls Laridae
Europe, Asia, North America, North Africa
© NHPA/PHOTOSHOT, JORDI BAS CASAA

FURTHER READING

Alcock, John. *Animal Behavior: An Evolutionary Approach*. 9th ed. Massachusetts: Sinauer Associates, Inc., 2009.

Bonner, John Tyler. *Why Size Matters: From Bacteria to Blue Whales*. Princeton: Princeton University Press, 2006.

Darwin, Charles. *On the Origin of Species*. Rev. ed. New York: Oxford University Press, 2008. First published 1859 by John Murray.

———. *The Descent of Man*. New York: Penguin Group (USA) Inc., 2004. First published 1871 by John Murray.

Foster, Russell & Leon Kreitzman. *Rhythms of Life*. New Haven: Yale University Press, 2005.

Smith, John Maynard. *The Theory of Evolution*. 3rd ed. Cambridge: Cambridge University Press, 2008.

Ridley, Mark. *Evolution*. New York: Oxford University Press, 2004.

A FIREFLY BOOK

Published by Firefly Books Ltd. 2009
Copyright © 2008 Magnus Edizione

First printing

Publisher Cataloging-in-Publication Data (U.S.)
Minelli, Alessandro.
Surviving : how animals adapt to their environments / Alessandro Minelli and Maria Pia Mannucci
Originally published: Sopravvivenza, Italy: Magnus Edizioni, 2008.
[200] p. : col. photos. ;   cm.
Includes bibliographical references.
Summary: Examines the many ways animals survive or cope with different aspects of their lives, such as hunger, love, enemy attacks, extreme conditions and change.
ISBN-13: 978-1-55407-520-1
ISBN-10: 1-55407-520-3
1. Animals--Adaptation. I. Mannucci, Maria Pia. II. Title.
591.4 dc22   QP82.M563   2009

A record for this book is available from Library and Archives Canada

Published in the United States by
Firefly Books (U.S.) Inc.
P.O. Box 1338, Ellicott Station
Buffalo, New York 14205

Published in Canada by
Firefly Books Ltd.
66 Leek Crescent
Richmond Hill, Ontario L4B 1H1

Cover Design: Kate Panek

Printed in China